SHAPING THE DIGITAL DISSERTATION

Shaping
the Digital Dissertation

Knowledge Production
in the Arts and Humanities

Edited by Virginia Kuhn
and Anke Finger

https://www.openbookpublishers.com

© 2021 Virginia Kuhn and Anke Finger (eds). Copyright of individual chapters is maintained by the chapters' authors.

ISBN Paperback: 9781800640986
ISBN Hardback: 9781800640993
ISBN Digital (PDF): 9781800641006
ISBN Digital ebook (epub): 9781800641013
ISBN Digital ebook (mobi): 9781800641020
ISBN XML: 9781800641037
DOI: 10.11647/OBP.0239

Cover image: Erda Estremera on Unsplash, https://unsplash.com/photos/eMX1aIAp9Nw. Cover design by Anna Gatti.

Contents

Acknowledgements

Virginia Kuhn began this collection several years before its publication, enlisting the help of Kathie Gosset as co-editor. Both realized their time was limited, but they also felt the need for this collection quite strongly. Just when Kathie's schedule made her continuing involvement untenable, Virginia was diagnosed with stage 2 breast cancer making the whole enterprise seem doomed to remain incomplete. With Virginia's strong recovery, the addition of Anke Finger, one of the collection's authors, as co-editor, the patience of our contributors and the good will of this Press, we are proud to see this collection come to fruition.

Over the course of shaping this project into a book, a number of colleagues and institutions have been immensely supportive. Virginia would like to thank Kathie Gossett, the value of whose early work molding the volume cannot be overstated. Her chapter, co-authored with Liza Potts, is a vital addition to this collection. Virginia would also like to thank her colleagues and graduate students whose careful thinking, pedagogical excellence and collegiality have provided a sounding board and intellectual home for the many years since she defended her own digital dissertation in 2005.

Anke Finger would like to thank the Humanities Institute at the University of Connecticut for their generous support of this book project. She is immensely grateful to all graduate students who have embarked on explorations within digital scholarship, and who worked towards making Digital Humanities and Media Studies a lasting initiative, together with many colleagues within and beyond UConn. The creativity and curiosity have been and continue to be enormously inspiring; the same is the case for everyone involved with this volume, the contributors and, most particularly, Virginia Kuhn. May the final product inspire more cutting-edge work in the future.

We would also like to thank Alessandra Tosi and the team from Open Book Publishers, who made this process a transparent and ethical one. Their professionalism and warmth demonstrate that excellence and humanity are not mutually exclusive; rather they serve each other. We extend our particular gratitude to Adèle Kreager for her careful, insightful and prompt editing; Adèle made the process quite painless during what is an unprecedented moment in human history.

Contributor Biographies

Dr. **Nicky Agate** is the Snyder-Granader Assistant University Librarian for Research Data & Digital Scholarship at the University of Pennsylvania and a co-PI on the HuMetricsHSS initiative, which promotes a values-based, process-oriented approach to evaluative decision making in the academy. She serves on the steering committee of the Force Scholarly Communication Institute and the editorial board of the Journal of Librarianship and Scholarly Communication.

Cécile Armand is currently a Postdoctoral researcher in the European Research Council (ERC)-funded project "Elites, Networks and Power in modern China" (Aix-Marseille University). Prior to assuming this position, she obtained a two-year postdoctoral fellowship from the Foundation for International Scholarly Exchange (2018–2020) and the Andrew Mellon Foundation as part of the DHAsia Program at Stanford University (2017–2018). She completed her PhD in History at the Ecole Normale Supérieure of Lyon (ENS Lyon, France) in June 2017. Her dissertation dealt with a spatial history of advertising in modern Shanghai (1905–1949). She also led an interdisciplinary junior research lab devoted to digital humanities at ENS Lyon.

Cheryl E. Ball directs the Digital Publishing Collaborative and the Vega publishing project at Wayne State University. She is executive director for the Council of Editors of Learned Journals, serves as the Editor-in-Chief for the Library Publishing Curriculum, and is editor of Kairos, the longest continuously running scholarly multimedia journal in the world. See http://ceball.com for her full CV.

Allison C. Belan is the Director for Strategic Innovation at Duke University Press. Allison leads critical strategic initiatives and drives the development and execution of the organization's strategic plan.

She manages the Press's IT, business systems and digital content teams. Prior to assuming this role, Allison worked at Duke University Press in a variety of roles, including Journal Production Manager and Director for Digital Publishing.

Sarah-Mai Dang is Principal Investigator of the BMBF research group "Aesthetics of Access. Visualizing Research Data on Women in Film History" (DAVIDF) at the Institute of Media Studies, Philipps University Marburg, Germany. Additionally, she is project leader of the international DFG research network "New Directions in Film Historiography". With a doctoral degree in Film Studies from Freie Universität Berlin and a Master of Arts from the University of Michigan, her current research and teaching focus on digital data visualization and the production and dissemination of scholarly knowledge. Based on her research interests, she founded the hybrid self-publishing project oa books, where she published her dissertation on aesthetic experience, feminist theory and chick flicks and blogs about academic publishing. She also initiated the Open Media Studies Blog and the Open Media Studies working group.

Anke Finger is Professor of German and Media Studies and Comparative Literary and Cultural Studies at the University of Connecticut. She has published widely in Modernism, Media Studies, and Intercultural Communication. She is the co-founder and co-editor (2005–2015) of the multilingual, peer reviewed, open access journal *Flusser Studies*. From 2016 to 2019 Anke Finger served as the inaugural director of the Digital Humanities and Media Studies Initiative at the Humanities Institute; she also co-founded the German Studies Association's Network on Digital Humanities and served as co-director from 2017–19. She founded the NEHC-DH network, affiliated with the New England Humanities Consortium; and she co-founded the CTDH network.

Kathleen Fitzpatrick is Director of Digital Humanities and Professor of English at Michigan State University. Prior to assuming this role in 2017, she served as Associate Executive Director and Director of Scholarly Communication of the Modern Language Association, where she was Managing Editor of PMLA and other MLA publications, as well as overseeing the development of the MLA Handbook. She is project director of Humanities Commons, and co-founder of the digital scholarly network MediaCommons. She currently serves as the chair of the board of directors of the Council on Library and Information Resources.

Erin Rose Glass is a researcher and consultant whose works focuses on education and ethics in digital environments. She is co-founder of the online learning community Ethical EdTech, co-founder of Social Paper, a networked platform for student writing and feedback, and founder of KNIT, a non-commercial digital commons for higher education in San Diego. She currently works as a Senior Developer Educator at Digital Ocean.

Kathie Gossett received her PhD from the Center for Writing Studies at the University of Illinois, Champaign-Urbana. She is currently a member of the faculty in the University Writing Program and an affiliate member of the Digital Humanities Institute at the University of California, Davis. Kathie has published in journals such as *Kairos: Rhetoric, Technology, and Pedagogy, Computers and Composition Online* and *MediaCommons*. She has also contributed multiple book chapters to edited collections, and has led several digital development projects. Before returning to graduate school, she worked in the information technology sector as a project manager, systems designer, user experience specialist, web designer/architect and technical communicator.

Virginia Kuhn is Professor of Cinema and Associate Director of the Institute for Multimedia Literacy in the Division of Media Arts + Practice at the University of Southern California's School of Cinematic Arts. In 2005, she successfully defended one of the first born-digital dissertations in the United States, challenging archiving and copyright conventions. Kuhn has published (with Vicki Callahan) an anthology titled *Future Texts: Subversive Performance and Feminist Bodies* (Parlor Press, 2016) and has edited several digital anthologies including: 'The Video Essay: An Emergent Taxonomy of Cinematic Writing' (*The Cine-Files*, 2017) with Vicki Callahan; *MoMLA: From Panel to Gallery* (*Kairos*, 2013) with Victor Vitanza; and *From Gallery to Webtext: A Multimodal Anthology* (*Kairos*, 2008) with Victor Vitanza.

Anthony Masure is the Head of Research at HEAD—Genève (HES-SO). He is a graduate of the École Nationale Supérieure (ENS) of Paris-Saclay, where he studied design. He is a research fellow of the lab

LLA-CRÉATIS at Toulouse—Jean Jaurès university. His research focuses on social, political and aesthetic implications of digital technologies. He cofounded the research journals *Réel-Virtuel* and *Back Office*. His essay *Design et humanités numériques* (*Design and digital humanities*) was published in 2017 by Les Éditions B42 (Paris). Website: http://www.anthonymasure.com

Monica McCormick is Associate University Librarian for Publishing, Preservation, Research and Digital Access at the University of Delaware Library, Museums and Press, where she leads IT, Digital Scholarship & Publishing, Digital Collections & Preservation, and the University of Delaware Press. The first half of her career was in university press publishing, mostly as acquiring editor for history and ethnic studies at the University of California Press. She received her MSLS at the University of North Carolina, Chapel Hill.

Joshua Neds-Fox is Coordinator for Digital Publishing at Wayne State University Libraries. He helps guide the development and direction of the Libraries' digital collections infrastructure and institutional repository, and collaborates with Wayne State UP to house their online journals. He co-edited the Library Publishing Coalition's *Ethical Framework for Library Publishing*, and serves on the Editorial Board of the Library Publishing Curriculum.

Thomas Oates is an Associate Professor with the Department of American Studies and the School of Journalism and Mass Communication at the University of Iowa. His research focuses on sports media and critical/cultural studies.

Lena Redman (**aka Elena Petrov**) completed her PhD at the Faculty of Education, Monash University, Melbourne, Australia. During her doctoral study, she developed the methodology of multimodal cinematic bricolage as an approach for knowledge construction. Redman generated a pedagogical model of Ripples, which integrates the agentic values of the individual with the concept of privatized tools of knowing. She is the author of *Knowing with New Media: A Multimodal Approach For Learning*. Detailed information for the Ripples pedagogy can be found at http://www.ripplespedagogy.com

Liza Potts is a Professor at Michigan State University where she is the Director of WIDE Research and the Co-Founder of the Experience Architecture program. Her research interests include networked participatory culture, social user experience, and digital rhetoric. She has published three books and over 70 publications focused on disaster response, user experience, and participatory memory, as well as several digital projects from community archives to leading the Sherlockian. net team. Her professional experience includes working for startups, Microsoft, and design consultancies..

Celeste Sharpe is faculty in the History and Political Science department at Normandale Community College. She earned her PhD in History from George Mason University in Fall 2016 and successfully defended the department's first born-digital dissertation. Previously, she was Interim Director for Academic Technology at Carleton College, a Penn Predoctoral Fellow for Excellence Through Diversity at the University of Pennsylvania. Additionally, she worked at the Roy Rosenzweig Center for History and New Media on a number of history education and public history projects.

Lisa Tagliaferri is an interdisciplinary scholar in literature and the computer sciences. Currently, she is the Kress Digital Humanities Fellow at Harvard University's Center for Italian Renaissance Studies. Previously, she was a postdoctoral researcher at MIT in the Digital Humanities program. Her widely downloaded programming book, *How to Code in Python*, has been adopted in classrooms as an open educational resource. She holds a PhD in Comparative Literature and Renaissance Studies from the City University of New York and an MSc in Computer Science from the University of London.

Katherine Walden is an Assistant Teaching Professor in the Department of American Studies at the University of Notre Dame. She holds concurrent appointments in Gender Studies and Computer Science and Engineering. Her research and teaching focus on critical approaches sport, data, and digital technology.

Christopher Williams is a Senior Scientist/Postdoctoral Fellow at the Doctoral School for Artistic Research, University of Music and Performing Arts Graz. He holds a PhD from Leiden University and a BA from the University of California, San Diego. Williams makes, curates, and researches (mostly) experimental music. As a composer and contrabassist, Williams' work runs the gamut from chamber music, improvisation, and radio art to collaborations with dancers, sound artists, and visual artists. He co-curates the Berlin concert series KONTRAKLANG. His artistic research focuses on improvisation and notation; it takes the forms of both conventional academic publications and practice-based multimedia projects. www.christopherisnow.com

Introduction

Shedding Light on the Process of Digital Knowledge Production

Anke Finger and Virginia Kuhn

While digital dissertations have been around for many years, the processes by which they are defined, created and defended remain something of a mystery. Is an interactive PDF significantly different from its paper-based counterpart? What specific possibilities can a digitally networked environment open up that would be impossible in print? How are dissertation committees able to gauge the quality of natively digital work? What support systems and workflows do students need to complete these types of projects? How do digital projects change the ways faculty members advise doctoral students? What are the implications of born-digital dissertations for career choices, hiring potential and work beyond the academy?

Shaping the Digital Dissertation: Knowledge Production in the Arts and Humanities addresses these questions in a book whose chapters explore the larger implications of digital scholarship across institutional, geographic and disciplinary divides. Indeed, the issues are all the more pressing as universities have moved online in response to the pandemic, revealing the need for both greater epistemological experimentation and more creative pedagogy. This raises even more questions about the future of scholarship. The book consists of two sections: the first, written by senior scholars, uses jargon-free language to tackle some conceptual concerns around directing and assessing dissertations, as well as doctoral education more broadly. The second section consists

   https://doi.org/10.11647/OBP.0239.16

of nine narratives written by those who have successfully created and defended a natively digital dissertation. These narratives were carefully selected for their ability to represent a diverse set of disciplinary and institutional settings. Within these specialized contexts, however, the chapters also serve as case studies that address common themes faced by doctoral students as well as their advisors.

The impetus for this collection arose at the inaugural meeting of the Digital Humanities and Videographic Criticism Scholarly Interest Group of the Society of Cinema and Media Studies (SCMS) in 2017. A graduate student asked whether the group might consider gathering information regarding digital doctoral dissertations. One of this collection's editors, Virginia Kuhn, defended a natively-digital, media-rich dissertation in 2005, and had supported several others in the intervening years as well as written a lead article on the topic in *Academe*, the magazine of the American Association of University Professors in 2013. Given her long-time involvement with generating digital scholarly work, she was rather surprised by this request. In the discussion that followed, however, it became clear that some sort of database was very much needed, as was a collection of more detailed essays about the trials and tribulations of creating a doctoral thesis digitally. Indeed, although digital dissertations—by which we mean those that are not just traditional, word-based texts that are archived digitally—have been around for decades, there remains confusion about the processes that go into creating and assessing them. And this confusion is perhaps most keenly felt among doctoral advisors and committees, even as some of the more experimental work, such as A.D. Carson's dissertation which took the form of a 34-track rap album, was accepted for publication in 2020 by the University of Michigan Press.[1] These cases have been too few and far between to see them as a trend.

This collection then, is written as much for that constituency—advisors, administrators, graduate school representatives—as it is for

---

1   Carson created and defended his dissertation at Clemson University in 2017 under the direction of Victor Vitanza, the pioneering rhetorician who was also on Virginia Kuhn's 2005 dissertation committee. The University of Michigan published it in 2020 (A.D. Carson, *I Used to Love to Dream* (Ann Arbor: University of Michigan Press, 2020), https://doi.org/10.3998/mpub.11738372). See Colleen Flaherty, 'Scholarly Rap', *Inside Higher Ed* (October 5, 2020), https://www.insidehighered.com/news/2020/10/05/university-michigan-press-releases-first-rap-album-academic-publisher

current graduate students contemplating the form that their thesis may take. As such, we felt that the format must be accessible to this group via a printed book, one which also carries the gravitas of a prominent press, if it were to be taken seriously, shared widely, and become useful. To this end, the collection of essays we have assembled represents several disciplines and institutions, showcasing multiple approaches to doctoral research and scholarship. These differing approaches force us to consider what we mean when we speak of the 'digital dissertation': is it word-based but disseminated online? Is it multimodal? Is it a thesis that takes various (media) forms? One with a digital companion? These are vital considerations if doctoral education is to retain its standards of excellence while also remaining relevant to the larger world and if it is to embrace the affordances and communicative advantages of different media for the dissemination of new scholarship. This collection frames digital dissertations as those that could not be accomplished if done on paper; it means they use digital modalities beyond just words (multimodal), or they take advantage of the capabilities of a digitally networked world.

The Current State of Digital Scholarship

In 2006, the Modern Language Association issued a report on 'Evaluating Scholarship for Tenure and Promotion' listing twenty recommendations to address a perceived crisis in producing scholarship, with monographs maintained as the gold standard for tenure along with pressure for an increased volume of publications. While identifying types of scholarship that should be recognized, the report emphasizes as particularly 'troubling the state of evaluation of digital scholarship [...]: 40.8% of departments in doctorate-granting institutions report no experience evaluating refereed articles in electronic format, and 65.7% report no experience evaluating monographs in electronic format'.[2] Clearly, the definition of digital scholarship encompassed written work in digital form, not multimodal work or quantitative digital humanities. In fact, right around this time, 2005, Anke was advised against starting an open

---

2   Modern Language Association, *Report of the MLA Task Force on Evaluating Scholarship for Tenure and Promotion* (New York: MLA, 2007), p. 11, https://www.mla.org/content/download/3362/81802/taskforcereport0608.pdf

access, peer-reviewed, online journal—*Flusser Studies*—for fear of such work not counting for tenure. The journal is in its second decade, and it did count towards tenure, although not significantly. As Anke was up for promotion to full professor in 2016, she wondered whether evaluative measures at her institution had changed. Not much, as it turns out—her video essay on Vilém Flusser and multimodal thinking featured only marginally in her review letters, despite it garnering over 12,000 views on Vimeo, a readership many of us can only dream of for our written academic work.

Clearly, we have come a long way with many professional associations, including the Modern Language Association, the American Historical Association, the College Art Association and the Association for Computers and the Humanities now including digital scholarship worth counting towards PhD degrees and tenure and promotion. Contributions such as Jennifer Edmond's edited volume on *Digital Technology and the Practices of Humanities Research* help to broaden both the discussion of technology's impact on research and changing practices in the various humanities disciplines.[3] However, while there are guidelines for general evaluative measures issued by all, there are few if any specific parameters for advisors as intellectual chaperones or co-conspirators in the process of supporting a graduate student doing work that differs significantly from traditional dissertating structures and approaches. Certainly, institutions of higher learning should not abandon standards, but they must also acknowledge the fact that these standards are not immutable, nor ideologically neutral. Indeed, Yale University's first doctoral dissertation, created in 1861, was hand written on six sheets of paper. Dissertations quickly grew longer as inexpensive paper, typewriters and carbon paper became available.[4] This is a good reminder of the ways that academic outputs shift in light of the technologies of their production: the typewriter, the mainframe computer, the personal computer and, finally, the networked computer or mobile device.

Given the centrality of media affordances for knowledge production in general, one of the most important roles for those in humanities disciplines,

---

3   Jennifer Edmond, ed., *Digital Technology and the Practices of Humanities Research* (Cambridge, UK: Open Book Publishers, 2019), https://doi.org/10.11647/OBP.0192

4   Richard Andrews et al., *The Sage Handbook of Digital Dissertations and Theses* (London: SAGE Publications, 2012), p. 7.

we believe, is the cultural critique they can offer. Few other disciplines are able to comment on structural imbalances, institutional inequities, and outdated policies. By extension, few disciplines can offer deep readings of changes in knowledge production and their facilitating, accompanying or adjacent technologies. Perhaps more than the social sciences, which tend to focus on researching current structures and institutions, humanists can be activists and weigh in on cultural issues, suggesting changes for remedying the types of inequities and shortcomings we see. We should also be weighing in on matters of public interest, including career diversity for PhDs in the arts and humanities. Thus, this critique includes the culture of technological innovation and adoption. While technologists imagine things that *could* be, we imagine what *should* be.

We have done a good job of sequestering ourselves in our ivory towers, leaving ourselves vulnerable to misrepresentation by anti-intellectual forces. Indeed, if Pew research polls are to be believed, there has never been a moment when higher education, at least in the US, has been so little supported by the public. Academics can bridge this divide via their teaching since we reach so many students, who are, after all, future members of the general public. A text that has been hugely influential on Virginia's own pedagogy is bell hooks's *Teaching to Transgress*,[5] a book that includes an extended conversation with Paolo Freire, best known for his championing of critical pedagogy. Henry Giroux is also a continual source of inspiration regarding critical pedagogy but new voices are emerging: in *Radical Hope*, Kevin Gannon calls for a far more focused attention to teaching.[6] As we both have long argued, our relationship to students should not be adversarial but one of advocacy, advocacy in the spirit of 'generous thinking', as presented by Kathleen Fitzpatrick in a recent book,[7] but also by noting Jessie Daniels's and Polly Thistlethwaite's explication of what it means to be 'a scholar in the digital era'—namely by impacting and communicating with the public.[8]

5   bell hooks, *Teaching to Transgress: Education as the Practice of Freedom* (New York: Routledge, 1994).

6   Kevin Gannon, *Radical Hope: A Teaching Manifesto* (Morgantown: West Virginia University Press, 2020), https://doi.org/10.7551/mitpress/11840.003.0001

7   Kathleen Fitzpatrick, *Generous Thinking: A Radical Approach to Saving the University* (Baltimore: The University of Johns Hopkins Press, 2019).

8   Jessie Daniels, and Polly Thistlethwaite, *Being a Scholar in the Digital Era. Transforming Scholarly Practice for the Public Good* (Chicago: Policy Press, 2016), https://doi.org/10.1332/policypress/9781447329251.001.0001

One long-held apprehension about the public nature of digital technologies concerns both copyright and intellectual property. In the latter case, people worry that if they put their ideas online, they will be robbed of them; in the former case, people are nervous about using any type of sound or video fearing they will be accused of copyright infringement. These issues are actually two sides of the same coin and, in both cases, the answer hinges on citation practices. The best way to establish your authorship of an expression of an idea is to have a record of it—in other words, to put it online. Likewise, the best way to demonstrate your awareness of others' intellectual property (IP)— whether that IP resides in words, images or sounds—is to cite your sources.

Another ongoing concern has to do with the conflation of the words 'public' and 'published' and the prevailing idea that simply putting something online is the same as publishing it. The corollary notion is that if something is online, it is no longer of interest to publishers since it has already been 'published'. However, the jurying function that a publisher fulfills is key to any publication and, in fact, in several experiments with online peer review before the publication of a book, publishers found that the online version did not limit book sales.[9] Much of the bias against online publishing likely stems from these misguided notions that were rampant in the early days of the internet and will certainly persist if they are not examined by the academic community. Such bias, we hold, not only impedes the sharing of new ideas and innovative scholarship because it is deemed a hazard, it also blocks vital dialogue between two cultures that have artificially distanced themselves over time, academia and the public commons.

According to Marissa Parham, 'in 2018 digital work is still often an unreasonably risky pursuit for many faculty, staff and students', noting that one must also produce traditional scholarship or have a record of non-digital publication before this risk abates.[10] In fact, many institutions

---

9   The Institute for the Future of the Book hosted many of these experiments, the first
     of which was done with McKenzie Wark's G3mer Theory, already under contract
     with Harvard University Press, the draft of the text was open for commentary
     online, and many of the comments made it into the final (printed) book. See
     https://futureofthebook.org/mckenziewark/
10   Marissa Parham, 'Ninety-Nine Problems: Assessment, Inclusion, and Other Old-
     New Problems', *American Quarterly*, 70.3 (2018), 677–84 (at 677–78), https://doi.
     org/10.1353/aq.2018.0052

issue indefinite guidelines, if any, for innovation and change that will be rewarded. Parham, for example, emphasizes that digital scholarship evaluation processes, if they are formalized, can reveal 'assessment as a site of miscommunication and unacknowledged institutional disinterest in transformation'.[11] If innovation and transformation are not part of the evaluative process, how can they be rewarded?

We think we can do better at communicating the value of born-digital scholarship and at merging both hermeneutic and heuristic practices in the humanities. When Anke asked the chair of her Promotion and Tenure Review committee what would help the members to evaluate digital scholarship projects, he mentioned the necessity of training workshops, and he suggested two items, followed by a question mark: 'A rubric providing a comparative basis for digital works and, perhaps, a comparative basis for digital and non-digital works?' He knows we have an intercultural communication problem because we are trying to compare apples to oranges. In Anke's mind, scholarship evaluators in the humanities are not print-centric by choice or sheer obstinacy— they/we/you are print-centric by habituation and acculturation and subscribe to scholarly value systems that seek to maintain rigorous quality control, a highly-charged value from an emic perspective. How do we change these habits to allow for innovation in both form and content? The dissertation, more so than any other academic genre, is the first step towards intellectual innovation where the new hypothesis or question receives room for experimentation: why has it been so difficult to establish this genre as the best laboratory or playground to test an innovative thinker's mettle, to provide a relatively secure ground for taking off in new directions?

I. Issues in Digital Scholarship and Doctoral Education

The first section comprises six chapters by nationally and internationally recognized scholars who have either contributed to, shaped or started the conversation about born-digital dissertations and digital scholarship in general. In this section, the authors speak to the variety of changes in scholarship, changes that include moving beyond a traditional and traditionally secluded discourse and knowledge mediation; to the changes

---

11   Ibid., 679.

in advising PhD candidates who are expecting a variety of knowledge designs commensurate with their everyday communicative experiences; and to a variety of infrastructural, strategic, and organizational issues universities face when pursuing educational and research goals for the twenty-first century. The audience for whom this portion of the project is intended, doctoral advisors and dissertation/thesis committees in the arts and humanities, are these authors' peers. As such, the six chapters speak directly to those in charge of initiating and navigating the aforementioned changes, for example, by applying the second section's narratives productively such that the larger discussion—for each PhD-granting department—may be tied to routinizing approaches and practices. These contributions may also inspire more broadly conceived discussions within graduate schools and upper administration units to facilitate structures supporting digital dissertations in general. The section concludes with a step-by-step guide to establishing and carrying out digital scholarship including best practices for discoverability and preservation.

II. Shaping the Digital Dissertation in Action

The second section comprises nine chapters composed by PhD students in the arts and humanities, though all are informed by different disciplinary and geographical/cultural vantage points. These narratives—examples of dissertating experiences and outcomes that speak to the variety of options in both form and content—present blueprints for doctoral advisors and dissertation/thesis committees as well as for PhD students just embarking on their dissertation and who seek peers or mentors outside of traditional scholarly support systems.

The topics addressed in these nine chapters include modes of production (impact, copyright and ethics); multimodal scholarship (adding sound, image, non-linear narrative and interactivity); dissemination (for a globally networked society, including audience engagement); and versioning (multiple versions of the same dissertation for different audiences or access to different formats). Each author reflects not only on their individual challenges with digital scholarship as a burgeoning and necessary approach to their academic work, they also present, in accessible language, the processes of production and dissemination unique to their outcomes. All narratives raise issues pivotal to academic work in the twenty-first century: how does knowledge production (traditionally confined within the intellectual walls of peer review, strictly structured, linear communication and costly print publications) engage with media beyond print, engage the public, and engage in epistemological innovation? The chapters in the second section are strategically placed in order to show the range of possibilities for scholarship in a globally networked world. The early chapters make use of the networking potential in order to reach a wide audience beyond academia. These are largely word based. The middle chapters are more hybrid in nature, often requiring several versions of the same dissertation as appropriate to various rhetorical situations and formats. The final chapters make use of the multimodal capabilities offered by digital technologies; they incorporate the textual as well as the aural and the visual. These dissertations are especially provocative in that they challenge the primacy of verbal language as the only and best form of argument.

The combination of a book about the complexities of digital scholarship (Section I) within which authors also speak about the process of planning, composing and defending their digital dissertations (Section II), makes this project not only unique but, we hope, generally useful to its intended readership: it offers a wide variety of evidence about the value of and need for digital scholarship at the doctoral level. Indeed, digital scholarship in the arts and humanities, we argue, mirrors the media landscapes available to researchers in the twenty-first century and broadens the variety of methodological approaches to innovative inquiry beyond traditional knowledge design.

The essays here enliven the conversation as they recount some of the historical and conceptual efforts carried out in the name of digital scholarship. Kathleen Fitzpatrick opens the collection with an analysis of the sudden isolation graduate students find themselves in during the dissertation process. In the humanities, she observes, graduate students are regularly habituated into an anxiety of intellectual independence whereby sharing ideas, collaboration and publishing work in progress is to be considered suspect and potentially diminishes its scholarly value. Digital scholarship, she argues, can eliminate or at least sideline such anxieties (and their untimeliness) by creating a participating public, testing ideas, interesting possible publishers early and creating a community of scholarship that, together with the support of PhD-granting institutions, endorses 'new kinds of open work'. Cheryl Ball, too, emphasizes the need for open work in the form of open access facilitation. Adding a historical view towards digital scholarship formats and highlighting the library's role in archival practices, she suggests that digital dissertations play a significant role in embodying the possibility of sharing scholarship publicly and that librarians are pivotal collaborators for any digital scholarship endeavor. Significantly, Ball also emphasizes the need for openness when evaluating digital dissertation forms: why not approach digital work 'on its own terms' in order to allow for 'radical scholarship'? Fitzpatrick's call for freeing the dissertating student from isolation and Ball's underscoring openness both in approach and access to digital scholarship is echoed by Virginia Kuhn who, for years, has honed a loosely established rubric, refined in collaboration with a group of students, with which to assess digital theses. Three areas, 'Conceptual Core, Research Component, Form + Content', each feature three additional foci that leave ample room for epistemological play and space beyond a traditionally alphabetized, linear text-only dissertation. For example, digital scholarship need not be 'thesis-driven prose'; instead, it can establish a 'controlling idea' presented in media other than text. Any kind of rubric or assessment measure, Kuhn warns, also requires a rethinking of review formats, however: annotation and feedback, too, will necessitate multimodal features such that radical scholarship and deep collaboration, to use Ball's and Fitzpatrick's terms, become part of evaluative considerations and feedback formats allow for non-linear, creative interruptions.

Outlining the trials and tribulations of archiving born-digital dissertations, Kathie Gossett and Liza Potts detail a study they have conducted over more than a decade, the ultimate goal being the formation of a persistent, searchable database of these projects. The results of a National Endowment for the Humanities funded workshop conducted with stakeholders from several academic institutions, Gossett and Potts note their work on establishing a network of like minded scholars for support when working in nontraditional formats. Anke Finger shores up this focus on form with an incisive argument about the shifting nature of the book as both a 'medium and artifact', and one which offers exciting possibilities with the affordances of the digital. However, academic institutions, Finger notes, have not kept pace with these new forms and this is due, in large part, to a lack of evaluative measures and experience in applying them, making it risky at best to embark upon a large-scale digital project. Using her experience as a PhD advisor and founding director of the Digital Humanities and Media Studies initiative at the University of Connecticut, Finger argues for support for digital literacy in humanities-based graduate education. Specifically, she argues that students need 'access to scholarly inquiry and research innovation beyond print', and this should come early in graduate education in order to provide the type of scaffolding needed if universities are seriously committed to digital scholarship. Rounding out this section is a collaboratively authored chapter by digital librarians, publishers and archivists, who have established a heuristic dubbed FICUS which stands for findable, impactful, citable, usable and sustainable. These will be widely applicable across disciplines, formats and topics.

The chapters in the second section provide precedents for future dissertating students, while also offering candid descriptions of the obstacles encountered. Forming a bridge between the two sections, chapter seven features a dissertation student, Katherine Walden and her advisor, Thomas Oates who describe the questions they contended with and the steps taken to create and defend Walden's interdisciplinary digital thesis project in the field of American Studies. While there are signs of the field's recent support for and of digital scholarship, they note, many questions remain. And since many of the obstacles to Walden's dissertation were logistical and administrative in nature, her dissertation became a springboard to a larger conversation among faculty at the University of Iowa. Walden and Oates argue for the power of a precedent, and their chapter joins the expanding catalogue of models, offering both conceptual and instrumental advice to future doctoral students as well as their advisors.

Cécile Armand extends the call for rethinking the nature of the dissertation and academic argument in general. In chapter eight, Armand describes a digital database she created as a companion to her dissertation in Chinese history. This companion allowed her to make use of primary source materials that are not typically considered in scholarly work; these include newspaper advertisements as well as 'professional handbooks, business materials, municipal archives (including correspondence, regulations and technical sketches), street photographs, and to a lesser extent, original maps and videos'. Although Armand's first concern was the creation of a permanent home for these materials, this database actually impacted the written portion of her dissertation project since it allowed her a spatial view of her subject, for instance, which opened up new insights. This is an excellent example of the ways that form impacts content and vice versa.

Sarah-Mai Dang, working from within the context of German academic parameters, questions a publication process that relies on economic structures often beyond the reach of the graduate and maintaining the 'symbolic capital of the book'. Instead, she chose to publish her research in four different formats, trying to undo a staid and costly convention that not only prevented affordable (for both author and reader) public dissemination, but also a speedy delivery of scholarship and access to an international audience. Simultaneously, as a media studies scholar, she turned this process into a research project, taking stock of data to measure impact.

The desire for and influence of a larger audience for academic work is extended by Erin Rose Glass as she describes the background and process of #SocialDiss, a project in which she posted drafts of her dissertation to a variety of online platforms for public review. Gauging the reviews and the many types of public and community engagement produced, Glass argues that academic writing, especially at the student level, would benefit from digital infrastructure, practices and incentives that emphasize collaboration and community building.

Lisa Tagliafari reinforces the need for academic work to reach a wider audience using her own dissertation as a case study. Not only does Tagliafari advocate for open source, hers was also the first chapter offered as a preprint to this collection, via the MIT's database. Her essay describes open source, open access and Creative Commons before offering suggestions for stakeholders to consider when navigating various levels of access. Anthony Masure's approach, while similar to Dang's in that he, too, sought to burst the limitations of print-only parameters common and expected in France, seeks to deepen the notion of his dissertation work's readability. Noting the technical hurdles of constantly updating a webpage, for example, he designs his PhD-thesis website by cleaning HTML code and without using a CMS, thus aiming for a 'true' version of his dissertation that, in fact, supersedes the version he submitted to obtain his degree. Ultimately, Masure leads us back to Tim Berners-Lee by advocating for sharing knowledge without borders and critically engaging with the potentially limiting affordances of specific media prescribed for knowledge production. Similarly dismissing the epistemological confines of traditional thesis composition software such as Word, Lena Redman (aka Elena Petrov) devises her own theory of multimodal creativity by analyzing what she calls 'deep remixability' and its interdependence with 'cinematic bricolage' as a research methodology. Her thesis, composed with InDesign and the Adobe Cloud, employs mnemonic material and autobiographical information to enhance what Redman calls feedback loops. These loops deepen the researcher's individualization of knowledge as her intellectual work merges with memory-work to allow for unique meaning-making processes and what Søren Brier has called 'cybernetics of human knowing'.[12]

If the digitally networked world provides the ability to author with images as well as a more open form of academic scholarship, it also raises concomitant ethical considerations around areas such as privacy and copyright. Celeste Tường Vy Sharpe confronted these issues in her own dissertation project completed in a department of History. Sharpe's research included extensive archival research of sensitive materials in her exploration of visual culture and disability. Given the topic, Sharpe found herself weighing the need for visual evidence with the ethics of exposing images culled from the March of Dimes. Finally, Christopher A. Williams explores the deeper layers of web design to discover the communicative potential of 'sticky web galleries' for the multimodal and broad public dissemination of improvisation in music. He describes in great detail the collaborative process necessary to design his thesis in WordPress, complete with paths and multimedia files that align with musical knowledge, beyond linear text. As a team, he and his collaborator arrive at a site that 'as a whole functions as a sort of meta-score for improvisers'. At the same time, the thesis becomes not only

---

12 Søren Brier, ed., *Cybernetics and Human Knowing: A Journal of Second Order Cybernetics, Autopoiesis and Cyber-Semiotics* (1992-present).

a milestone within a research path, it also turns into a resource for practitioners outside of the usually closed publication loop as a 'living meta-work.'

Together, these essays demonstrate that digital dissertations, and digital scholarship as such, not only have a rich history already, but that, as a form of knowledge production in the academy, they are established modes of inquiry. The many topics addressed, from a plethora of perspectives and knowledge-bases, speak to the timeliness of examining the dissertation as a genre or space where scholarly innovation should be permitted even more room and openness to utilize tools, approaches, and methods at the scholar's disposal. For any 'radical scholarship' or transformation of scholarly practice is ultimately also tied to the technical and media parameters embedded in the scholar's environment of production and these environments are now allowing for remarkably creative, communicative and visionary work both inside and outside of academe.

Bibliography

Andrews, Richard, et al., *The SAGE Handbook of Digital Dissertations and Theses* (London: SAGE Publications, 2012).

Brier, Søren, ed., *Cybernetics and Human Knowing: A Journal of Second Order Cybernetics, Autopoiesis and Cyber-Semiotics* (1992-present).

Carson, A.D., *I Used to Love to Dream* (Ann Arbor: University of Michigan Press, 2020), https://doi.org/10.3998/mpub.11738372

Daniels, Jessie, and Polly Thistlethwaite, *Being a Scholar in the Digital Era. Transforming Scholarly Practice for the Public Good* (Chicago: Policy Press, 2016), https://doi.org/10.1332/policypress/9781447329251.001.0001

Gannon, Kevin, *Radical Hope: A Teaching Manifesto* (Morgantown: West Virginia University Press, 2020), https://doi.org/10.7551/mitpress/11840.003.0001

Edmond, Jennifer, ed., *Digital Technology and the Practices of Humanities Research* (Cambridge, UK: Open Book Publishers, 2019), https://doi.org/10.11647/OBP.0192

Fitzpatrick, Kathleen, *Generous Thinking: A Radical Approach to Saving the University* (Baltimore: The University of Johns Hopkins Press, 2019).

Flaherty, Colleen, 'Scholarly Rap', *Inside Higher Ed* (October 5, 2020), https://www.insidehighered.com/news/2020/10/05/university-michigan-press-releases-first-rap-album-academic-publisher hooks, bell, *Teaching to Transgress: Education as the Practice of Freedom* (New York: Routledge, 1994).

Kuhn, Virginia, 'Embrace and Ambivalence', *Academe*, 99.1 (2013), 8–13, https://www.aaup.org/article/embrace-and-ambivalence#XoT2ldNKjXG

Modern Language Association, *Report of the MLA Task Force on Evaluating Scholarship for Tenure and Promotion* (New York: MLA, 2007), https://www.mla.org/content/download/3362/81802/taskforcereport0608.pdf

Parham, Marissa, 'Ninety-Nine Problems: Assessment, Inclusion, and Other Old-New Problems', *American Quarterly*, 70.3 (2018), 677–84, https://doi.org/10.1353/aq.2018.0052

SECTION I

ISSUES IN DIGITAL SCHOLARSHIP AND DOCTORAL EDUCATION

1. Dissertating in Public

Kathleen Fitzpatrick

The process of writing a dissertation is often an exercise in profound isolation. Having begun graduate school as part of a cohort, having been closely supervised and surrounded through the process of coursework and qualifying exams, you are suddenly released and left to your own independent devices. In fact, the dissertation is intended as a test of those independent devices: can you self-motivate, self-regulate, develop and maintain a schedule to keep your work moving forward? The process is meant to enable the candidate to develop the self-reliant habits of mind that will serve them throughout their career. But what this exercise in independence frequently produces is far more troubling: the candidate runs headlong into loneliness, self-questioning and imposter syndrome.

These isolation-driven anxieties and doubts are so much a part of academic thinking about the individual long-form research project that we might begin to see them as features rather than bugs: tests of one's scholarly mettle. In fact, the profession has long since selected for the ability to withstand such isolation; those who make it through go on to design and oversee programs that impose the same conditions that were imposed on them. And of course, much of the later work that will be done by scholars who successfully join the tenure track—and that will be assessed, again, by those who have succeeded on that track—requires the same isolation, and the same ability to withstand it. After a certain point, in fact, we crave it: we want nothing more than to close the door, shut out the world, and focus on our individual projects.

But the isolation that is built into the dissertation process often comes at a profound cost: in some cases, to the individual mental health

  https://doi.org/10.11647/OBP.0239.01

of the scholars themselves, but in many more cases, to the health of the larger scholarly community. Being thrown out on our own, left to fend for ourselves, teaches us that the most important work that we do—the work on which our most important evaluations depend—must be done alone. We are pulled away from the more collective aspects of academic life and persuaded instead that the only work that matters, the only work that deserves our attention, is our 'own'. The dissertation is one of the most crucial phases of the process through which scholars self-replicate, and when we select for independence we select against community. In encouraging scholars in formation to close the door, shut out other demands and focus inward, we undermine the potentials for connection, for collaboration, for collective action that foster a sense of scholarly work as contributing to a social rather than personal good. We reinforce the individualistic, competitive thinking that I have argued is eroding not only our relationships with one another on campus but also the relationships between institutions of higher education and the publics that we serve.

That for so many established scholars alternatives to the isolation of the dissertation process are literally unthinkable is precisely a sign that such isolation has taken on the status of ideology. We may never get far enough away from our 'every tub on its own bottom' assumptions to fully embrace, for instance, the possibilities of a collaborative dissertation, though that very impossibility creates an interesting thought problem. (Impossible why? What is the dissertation meant to do in preparing a candidate for a career? Are there aspects of the career, or indeed entire future careers that we can today only dimly imagine, that might be better served by the affordances of a team-based project?) Even if we accept the single-author requirement for the dissertation as a given, however—at least for now; we have, after all, begun to move away, if gradually, from the assumption that the dissertation must be strictly composed of linear, text-based argumentation and analysis—there are ways that candidates might be encouraged to work more communally and publicly on dissertation projects, ways that might help alleviate some of the isolation and the problems that it creates.

In fact, many candidates rely on writing groups for both support and accountability in the dissertation process. Such writing groups tend to be local and private, a small cluster of scholars banding together to help one another through. It is possible, however, that more support might be found through scholarly networks online, through taking the leap to work on the dissertation in public. Public work like this can take a number of forms: it could be a matter of blogging about the process, about the ideas and the problems uncovered in the course of its research and composition. It could include posting drafts of chapters, or pieces of chapters, for discussion. In either case, the author could use a blog-based platform to work through challenges, to get feedback, to think about the significance of the project, and to build a sense of the community to whom the project speaks.

No doubt the last paragraph has the potential to induce an anxious reaction or two in some readers. If deep collaboration remains all but unthinkable in some corners of our scholarly lives, making work publicly available before it is 'ready'—before it's been revised, reviewed and given a professional seal of approval—is nothing short of impossible. We worry about the dangers inherent in allowing less-than-perfect work to be seen, about the possibility of having our ideas appropriated, about interfering with future publication opportunities. These worries are real, but also misplaced; they develop out of the general cloud of anxiety that covers the dissertation process, and they are heightened by well-meaning colleagues and advisors who do not always understand the potential benefits of working in public, or the ways that concerns such as these can be managed.

For instance: a willingness to make the process of developing the dissertation visible can not only help improve the project at hand but can also support future work, both one's own and that of others. Allowing work that is not yet perfect to be read and commented on not only can make possible early feedback from peers that can help guide the project's development, but it can also shed light on an occult process. And that visibility can benefit not just future dissertation writers but also many of our students: the hidden nature of our writing process too often leads novice writers to assume that our publications spring fully-formed from our heads; allowing them to see some of the messiness of our own processes can give them an understanding of what 'professional' drafting and revision look like, as well as the confidence to try it for themselves. It can also model for others—and for ourselves— the importance of conversation in the writing process.

Moreover, making the process of developing the dissertation visible can also demonstrate its potential to connect with a future audience. Projects that are written, or written about, in openly accessible ways can be found by editors who might be interested in working toward future publication. They can be found by other scholars who might be putting together collaborative projects in the field—conference panels or edited volumes, for instance—in which the work might play a role. And they can be found by journalists writing in related areas who might be interested in including the work in that reporting.

That last point raises its own set of concerns, of course, as scholars have recently complained about the growing tendency of such reporters to cite their sources inadequately at best, making it appear that the ideas developed through lengthy scholarly research and analysis are a mere part of the reporter's thinking. This is one of the several forms of 'getting scooped' that dissertation writers often worry about; other such worries include the possibility of another, faster scholar appropriating and publishing the work. These fears are, alas, real; a dissertation is designed to make an original contribution to the field, but it takes sufficiently long to be completed that one might reasonably worry about someone else catching wind of the idea and getting to the finish line first. However, these fears thrive on secrecy, and plagiarists, thieves and other unethical types are only able to get away with what they have done when there is no evidence that they have done it. In fact, the best way to avoid having one's work scooped is precisely *not* keeping it hidden away, but rather posting about it early and often. In this way, the ideas—complete with time stamps—come to be publicly associated with you, and any improper use can be equally publicly proven.

Finally, writing in public raises concerns for many candidates about the future publication possibilities for their dissertations, and how its public availability might disrupt them. On the one hand, it is true that university presses want the right of first publication for projects, and that the prior publication of that project online might diminish their interest. But that statement leaves out a few crucial qualifiers. First, university presses do not generally publish dissertations. Rather, they publish books that *develop out* of dissertations, and the distance between those two is more significant than it might sound. There is a lot of rethinking and revising involved in transforming a document largely written for a committee—designed to demonstrate one's mastery of a field and often responding to the idiosyncratic interests of one's advisers—into one written for a larger public. As a result, making aspects of the dissertation openly available—including depositing it in an open-access repository—will not necessarily cause a press to pass on the basis that it has already been published. In fact, a project that has already drawn online interest, and that has demonstrated its author's ability to write for and engage with a larger public, may well be appealing to those presses as the basis for a book.

And that last point is a key one to focus on: engaging with a larger public and developing a trusted network of readers interested in the work you are doing is of crucial importance. It is the key to overcoming the isolation involved in long-form scholarly work and to getting your work into conversation with the work of others. It is the key, in fact, to building a more open, more transparent, more generous scholarly community, because not only will your own work benefit from the connections that working in public can provide, but in fact the entire scholarly community can benefit. By finding more ways to work together, and to show the processes of our work, we can begin to make a bit more visible—a bit more accessible—what it is that scholars *do*. And that, in turn, might give us the potential to invite a range of broader publics into that work, creating a richer sense of why scholarly work *matters*.

Having arrived at this conclusion, however, I need to issue a strong final caveat: if greater forms of public engagement, of collaboration, of openness and community are key goals for scholars today, working toward those goals must not be left to them alone. We must consider what needs to change at the institutional level in order to support this work. That is to say, the impetus to work in public, and the responsibility for transforming their work, cannot lie solely at the feet of graduate students. Faculty, advisors and administrators must consider the ways that our curricula, our departments and our institutions facilitate and reward new kinds of open work, enabling it to be as transformative as possible. Only through such careful alignment of our institutions' internal processes and reward structures with the deepest values we hope to espouse can we begin to contribute to the most humane, most generous purposes of higher education: developing and sharing knowledge in order to foster and sustain engaged, caring communities for us all.

2. Publication Models and Open Access

Cheryl E. Ball

I have been participating in informal academic discussions of digital dissertations since first hearing about them while I was an undergraduate student at Virginia Tech in the early-1990s. Tech has been a pioneer in electronic theses and dissertations (ETDs), initiating the Networked Digital Library of Theses and Dissertations (NDLTD) to showcase ETDs that primarily used the Adobe Portable Document Format (PDF) to deliver digital versions of print-like dissertations.[1] A few years later, in 2000, I deposited what would be the first digital thesis for my Master's institution, Virginia Commonwealth University—a hypertextual and media-rich collection of creative writing to satisfy the requirements of my Master of Fine Arts in poetry. The steps to convince the university to allow what would be considered a 'nontraditional' model of publication were not difficult, and I was grateful for that. A book of poetry was already nontraditional in many senses of research in the academy (although not to creative writers), but I didn't face too many obstacles—or, perhaps, the length of time that has passed has lessened the memories of those obstacles.

Before I even began writing my thesis and with the acknowledgement of my thesis advisor, who approached my ETD ambitions with a modicum of rigor combined with a healthy dose of 'Good luck with that', I started at the top of my list: I wrote to the university president (so precocious!) to ask for permission to do this work, since our peer schools in Virginia had already taken up the ETD mantle. He agreed and put me on a

---

1   See http://www.ndltd.org/about

    https://doi.org/10.11647/OBP.0239.02

university-wide ETD Task Force. The members of that task force—the graduate dean and several faculty from across the disciplines—didn't quite know what to make of a poet who wanted to create an interactive, multimedia thesis when they were focused on making their students' scientific research accessible online in PDF format, but they were willing to listen, and I made good use of their time in showing them multiple examples of electronic poetry and fiction as well as identifying scientific PDFs from the NDLTD that showcased interactive 3D and other media elements embedded within the print-like dissertations.

Next, I went to the preservation librarian, who would ultimately be responsible for putting my ETD on a literal shelf in the library stacks, and asked her what the archival possibilities might be for a thesis that could only be read from a CD-ROM. She was very accommodating, showing me examples from the performance arts that included CD-ROMs of orchestrations along with the sheet music of composing students. She also required that the abstract and table of contents for the ETD be printed for metadata purposes (a word, to be sure, that I had never heard of and would not start actively using for more than a decade). Writing an abstract for a poetry collection was weird, but a required part of the deposit template needed so the work could be included in ProQuest's Thesis and Dissertation Abstracts index. The table of contents I played with a bit, since the collection was nonlinear and built to have multiple reading paths. I used the then-named Macromedia Director, a multimedia design software for creating interactive CDs, to build the collection and Storyspace, a literary hypertext authoring program, to create the table of contents, because the latter could show the multiple reading paths that were available between the twelve sets of poems I included in the poetry cycle. There were exponential reading paths possible, so I chose to show the visual map (see Fig. 1) of those paths that Storyspace created as well as a list of three possible paths in multiple-choice form for the final, bound thesis. That form of the thesis contained twelve printed pages, including the signature page, and a foam core to house the CD case.

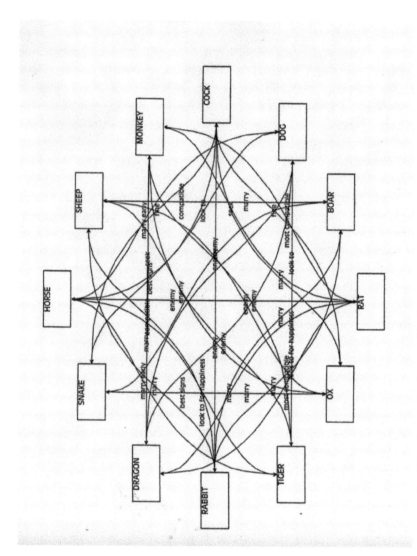

That was a long time ago. But some students are still having to navigate this process on their own and are also under the impression that this work is relatively new and they are not aware of the many, many precedents that have been set and resources that have become available over the last three decades of ETD work. This work is not new, even as it may be new to students and advisors and graduate deans. It is also not new to librarians, as my preservation librarian demonstrated in 2000. There are two points in that previous sentence that I want to discuss before returning to the idea of nontraditional models of ETDs. The first is that librarians are at the forefront of work with ETDs. The second is that significant work on ETDs began years before open access became a recognized term in scholarly circles.

Open Access and Why You Should Love Librarians

Librarians and archivists have had to figure out how to handle unusual scholarly texts and other materials at the point of collection and dissemination since long before any digital revolution hit our scholarly production workflows. Their jobs as information professionals have put them quietly (to most scholars) and squarely at the forefront of digital circulation and preservation issues in academia. That work is concomitant with open access as a default ethical value that librarians espouse—that is, open access, at its most fundamental level, is about making scholarship freely accessible to readers via the internet, and academic librarians promote access to knowledge at every turn. That is literally their jobs.

The term open access (OA) began widely circulating in 2001 after the December 2000 Budapest meeting of stakeholders interested in expanding the access of research beyond those who could most afford it. Research libraries have moved from solely being caretakers of scholars' print-like research at the end of its scholarly production lifecycle to being publishers and co-producers of OA research that takes advantage of the multiple technological platforms and genres available with Web-based circulation and preservation methods. OA scholarship has proliferated over the last twenty years thanks, in part, to the following technologies and genres that are possible with their use:

- institutional repositories (IRs), in which ETDs published by a university are typically archived, and faculty research from journals and other scholarly venues is re-posted, if copyright allows;

- open-access scholarly journals, including faculty and student-produced peer-reviewed venues that use either a university's IR or another academy-owned[2] open-source software platform to publish PDFs;

- digital humanities (DH) projects, as coordinated media- or data-intensive research projects created by librarians, between librarians and faculty members, or with librarians assisting faculty members who have digital projects that need sustainability (preservation and/or revision) plans the library can support;

- open educational resources (OER), which are collections of teaching materials put together in a coherent fashion, similar to a textbook, by an instructor to distribute for free to students.

By default, all of these project types facilitate open access publishing, which has been a mainstay in the sciences and in libraries since the advent of the World Wide Web in the mid-1990s.[3] This is not to say that the arts and humanities have not participated in this digital scholarly revolution—they have: From the first known, peer-reviewed journal that used email as a delivery platform (*Postmodern Culture*, c. 1990); to some of the earliest peer-reviewed literary arts criticism journals (*electronic book review*, c. 1995), electronic literature journals (*New River*, c. 1996), and digital rhetoric and pedagogy (*Kairos*, c. 1996); to some of the most recent advances in peer-reviewed publishing for the performing

---

2   **Academy-owned software** refers to (usually) open-source platforms that are developed by universities or other higher education institutions for use, usually for free but sometimes can incur customer service charges.

3   The humanities is not without its early innovators: the first known digital dissertation in the humanities is Christine Boese's 'The Ballad of The Internet Nutball: Chaining Rhetorical Visions from the Margins of the Margins to the Mainstream in the Xenaverse', which she defended in 1998 and wrote entirely in HTML with embedded images and exploratory navigational paths. There are several other early humanities examples, including Virginia Kuhn's 2005 highly visual dissertation, 'Ways of Composing: Visual Literacy in the Digital Age', authored in Sophie.

arts, including *The Journal for Artistic Research* (c. 2010) and the related multimedia repository, the *Research Catalogue*. Each of these venues has a different open-access business model (with the exception of *Postmodern Culture*, which is no longer open access). Yes, there are multiple business models for open access—the details of which are outside the scope of this essay—but all types of OA require that scholarly output be free to read, which exponentially expands a scholar's potential audience and engagement with publics (as Kathleen Fitzpatrick discusses in her essay in this collection). Yes, there are pitfalls and myths about OA that include a small percentage of predatory publishers who take advantage of the fear academics have in gaining and keeping employment—and shame on those publishers!—but detailing how to keep away from predatory vendors is also outside the scope of this essay as it's not immediately relevant to digital dissertations as the focus genre here.[4]

My point in detailing all the OA publication possibilities that are viable in a university setting is to strongly suggest that (1) digital dissertations have been published as OA texts longer than OA's existence and serve to bring a wider audience to one's research; so OA is not a thing to be feared, but to be embraced. And, (2) librarians are important collaborators for dissertators and their committees and can explain the OA environment in minute detail. A large research library might have an ETD librarian, a digital humanities librarian, an OER librarian, a 'scholcomm' (short for scholarly communications) librarian and maybe even a digital publishing librarian! The names may be different at every university, and a smaller PhD-granting university might have one person who fills all these roles (so be kind to them—they are definitely overworked!), but there will be someone in the library whose job it is to, at the very least, file your institution's dissertations with ProQuest (which is usually a requirement in the United States), so connect with that person early to ask for advice on creating a digital dissertation. Especially if that dissertation is expected to take a nontraditional form.

---

4   For a quick primer on avoiding predatory OA journals, use the Council of Editors of
     Learned Journals' 'Best Practices for Online Journal Editors' (2008), which provides
     a checklist for maintaining a reputable online journal, located at http://celj.org/
     resources/Documents/celj_best-practices-for-online-journals-REV.pdf

Publication Models for Digital Dissertations, or How Not to Pin People into Specific Genres

I started my academic career by publishing a collection of hypertextual poetry in an OA peer-reviewed journal that exclusively publishes scholarly multimedia texts. That poetry collection later became part of my digital thesis. I now edit that OA journal—*Kairos: A Journal of Rhetoric, Technology, and Pedagogy*, and have held that position for almost twenty years. (Yikes! And check us out at http://kairos.technorhetoric. net.) In that time, I have seen and participated in a lot of conversations about the shape of digital dissertations, and digital scholarship more generally in the humanities. As an extension of the research I did to prep for my MFA thesis and the webtexts I was editing for the journal, I wrote early on in my career a possible taxonomy for what we were then calling 'born-digital scholarship'—a name that some academics quickly realized was not that useful given how digitally embedded our scholarly practices had become, in our use of mundane and ubiquitous platforms like Microsoft Word. I was not then, nor am I now, excited to study scholarship that can primarily be represented by printing sheets of paper out and read via alphabetic text in a single, linear order. Instead, I have always been interested in how we might move away from 'digital scholarship' that is represented by print-like PDFs into more innovative, nonlinear and interactive media-driven forms.

Over nearly a decade, starting in the mid-2000s, the Modern Language Association's Committee on Information Technology slowly adopted and adapted Geoffrey Rockwell's wiki on digital scholarly genres for humanists, which included genres such as archives/ collections, TEI-based mark-ups of scholarly editions, and other projects that took advantage of hypertextual linking capabilities of the early Web.[5] I always took umbrage, however, that his list labeled hypermedia texts (what we might now call scholarly multimedia) as a 'nightmare' that were impossible to evaluate since they were never published in peer-reviewed venues (a patent falsity, even at the time he wrote the

---

5 Rockwell's wiki and the MLA's version of the revised guidelines are now both offline, but can be found in the 2011 print version of MLA's *Profession* in an article by Geoffrey Rockwell ('On the Evaluation of Digital Media as Scholarship', *Profession*, 1 (2011), 152–68, https://doi.org/10.1632/prof.2011.2011.1.152) in a special section on that topic.

list in the early to mid-2000s). While it is easy to take shots at a digital text that is no longer available, it is ironic that one of the main forms of digital scholarly production in the humanities has become those exact hypermedia genres, with many of the digital humanities projects being produced these days falling into the old-school category of hypermedia—that is, using the affordances of the Web (HTML with its capabilities of linking) to embed multimedia assets to create a holistic, multimodal meaning for a text.

Indeed since that time, I have witnessed many varieties of digital humanities genres that could fall into the category of hypermedia, in addition to the more stable genres of digitized collections, archives, and digital variorums. But that old classification of 'hypermedia models' vary in their generic representations as far and wide as there are authors to produce them and platforms with which to build them. That does not mean it is impossible to evaluate them in terms of quality as dissertators create their projects or post-PhD scholars produce similar projects as part of their research agendas. I have written several books and articles and held multiple week-long workshops on how to read, write and evaluate nontraditional, digital humanities projects including digital dissertations like the kinds represented in this book, and I can promise you—based on research that sampled over 1,000 webtexts produced over fifteen years—that the genres we encounter in digital, interactive, media-rich projects have not solidified.[6] And that is fine—and good, even! It just means that—like any text of any communicative mode we encounter as readers—we have to approach it on its own terms, figure out what genres it is using or remixing, hypothesize its narrative or rhetorical directions, follow our knowledge of gestalt to create meaning, and find closure on the text in the ways we know how to interpret. These are rhetorical acts of meaning-making that are necessary with any text we 'read'. For instance, in working with undergraduate and graduate students over a number of years to teach them how to author and evaluate scholarly multimedia texts, I asked them to create a list of key concepts they found useful to discuss sample digital media texts across a range of genres. We used some existing evaluative frameworks

6   See Cheryl E. Ball, 'The Shifting Genres of Scholarly Multimedia: Webtexts as Innovation', *The Journal of Media Innovations*, 3.2 (2016), 52–71, https://doi. org/10.5617/jmi.v3i2.2548

to start—including those that Virginia Kuhn has touched on in this book and written about extensively elsewhere as part of her work with the Institute for Multimedia Literacy. I then asked students to expand those frameworks to suit their own goals for authoring within the context of a specific assignment, which was to create an article-heft piece of scholarly multimedia, whereas dissertators might do the same with monograph-heft scholarly multimedia and similar digital humanities projects.[7] Some of the basic criteria touched on the relationship of a project's form to its content, and the innovative, creative or genre-defining or -bending work it does; the scholarly relevance, timeliness and appropriateness of a project given its suggested audience; and, of course, for scholarly genres, the validity and credibility of the research presented. Those are some broad rhetorical categories that can be added to with each piece of digital media, including digital dissertations, since they need to be evaluated within their own historical, technological, cultural and social framework, on their own terms, in relation to that moment and to the media and genres they use in that time. This is the same approach *Kairos* has taken in reviewing thousands of submissions for the last twenty-five years—a peer-review process, it should be noted, that is quite recursive with authors in the same way that advisors will be working with their advisees on dissertation projects.

Yes, there will always be texts that are difficult to parse because we have not encountered their like before. And, yes, there are ways to educate and mentor graduate students new to this composing process into understanding the rhetorical choices and genre conventions available to them so they're not just making shit up, or 'adding bells and whistles', as my thesis advisor and, later, a dean warned me not to do—a specious complaint to someone well enmeshed in this work, by the way, and hurtful to those just beginning their learning process. Dismissing the integral work of design and aesthetics, which are powerful meaning-making choices in their own right, in favor of some made-up notion of a purely rhetorical text is ridiculous and much derided in both art-based and non-art-based academic research areas including the fine

---

7 For an idea of how that framework plays out with some examples, see my article on 'Assessing Scholarly Multimedia: A Rhetorical Genre Studies Approach', *Technical Communication Quarterly*, 21.1 (2012), 61–77, https://doi.org/10.1080/10572252.201 2.626390

and performing arts, design, rhetoric, cultural studies and linguistics. Form *and* content both matter, and often simultaneously and with equal weight. So give students a chance before dismissing the kinds of radical scholarship their digital dissertations, in the form and content of digital humanities-type projects, might produce. This book showcases a wealth of contemporary examples and narratives for successful (and probably some not-so-successful) digital dissertations that can serve as additional models for those courageous enough to innovate in their digital research forms.

Bibliography

Ball, Cheryl E., 'Assessing Scholarly Multimedia: A Rhetorical Genre Studies Approach', *Technical Communication Quarterly*, 21.1 (2012), 61–77, https://doi.org/10.1080/10572252.2012.626390

Ball, Cheryl E., 'The Shifting Genres of Scholarly Multimedia: Webtexts As Innovation', *The Journal of Media Innovations*, 3.2 (2016), 52–71, https://doi.org/10.5617/jmi.v3i2.2548

Boese, Christine, 'The Ballad of The Internet Nutball: Chaining Rhetorical Visions from the Margins of the Margins to the Mainstream in the Xenaverse' (PhD dissertation, Rensselaer Polytechnic Institute, 1998), http://www.nutball.com/dissertation/

Council of Editors of Learned Journals, 'Best Practices for Online Journal Editors' (2008), http://celj.org/resources/Documents/celj_best-practices-for-online-journals-REV.pdf

Kuhn, Virginia, 'Ways of Composing: Visual Literacy in the Digital Age' (PhD Dissertation, UW-Milwaukee, 2005).

Rockwell, Geoffrey, 'On the Evaluation of Digital Media as Scholarship', *Profession*, 1 (2011), 152–68, https://doi.org/10.1632/prof.2011.2011.1.152

3. The Digital Monograph?

Key Issues in Evaluation

Virginia Kuhn

Faculty members who work in digital media or digital humanities should be prepared to make explicit the results, theoretical underpinnings, and intellectual rigor of their work.

MLA Guidelines for Tenure and Promotion, 2012.[1]

'This is a hobby. Don't let it distract you from the real work'. This well-intentioned warning issued by one of my graduate advisors came at the end of a workshop we'd just finished on digitizing video from tape. It was 2004 and YouTube did not yet exist but I was determined to get images into my work, sensing it would enrich my doctoral research significantly, even if I couldn't articulate exactly how and why at the time: on the one hand, my research was (and remains) engaged with issues of power and privilege. I investigate structural issues around race and gender—both very visual concerns—and the ways that they inform and are informed by the technologies used for communication and expression. This made it vital to actuate my argument *with* images. On the other hand, power differentials and structural inequities function best, and sometimes *only*, when they are invisible. In this light, any attempt to uncover power via the presumed literality or indexicality of visual media, by its very nature, undermines the complexities of power structures; the camera is not objective, nor are its photographic outputs comprehensive, and so the use of images must be carefully considered.

---

1 See https://www.mla.org/About-Us/Governance/Committees/Committee-Listings/ Professional-Issues/Committee-on-Information-Technology/Guidelines-for-Evaluating-Work-in-Digital-Humanities-and-Digital-Media

  https://doi.org/10.11647/OBP.0239.03

Ultimately, since my larger argument hinged on the premise that digital technologies are nearly as amenable to images as they are to words, it was compulsory to use image-based evidence and actually deploy this emergent visual language. And of course, nearly two decades later, images and video are so numerous online as to make the notion of their manipulation in a critical text an imperative.

Still, my advisor's warning reveals a key concern: how do graduate students—and academics more generally—decide how and where to focus their energy in a competitive environment that is built upon its members feeling they are never doing enough? How much time can we really afford to spend on learning a coding language, for instance, knowing we'll never become a developer even if we become a semi-decent programmer? And which programming languages or software applications are worth learning? Will the time spent translate into more insightful scholarship and how do we make that calculation? These decisions have real career implications, particularly since such an endeavor will likely not be seen as analogous to visiting an archive or spending time learning another natural language, the 'real work' of the humanities.

Perhaps more profoundly, however, my advisor's warning reveals the stubborn boundary between formal and conceptual elements in academic work: as such, the workshop we had completed which focused on manipulating the formal qualities of film via its digitization was seen as extraneous to its actual study and scholarship—the 'real' work of writing *about* film, not with it. Indeed, the form/content divide has held its own for hundreds of years, and to traverse it requires an explicit justification.

When creating a natively digital dissertation, this rationale is especially vital since there is little consensus on how to properly assess this work; as such, the chances of being penalized for these efforts are quite high. The digital text that carries the same intellectual heft of a traditional dissertation has not been identified with any precision, nor are even its general contours widely agreed upon. As the epigraph with which I opened suggests, and more than a decade of supervising dissertations confirms, more often than not, it continues to be the responsibility of the student to explain the 'results, theoretical underpinnings and the intellectual rigor' of their digital work. In what follows then, I suggest a rubric for evaluating born-digital scholarship—that which could not be done on paper—to help dissertation students articulate the merits of their work and, in the process, potentially educate their faculty advisors, or, at the very least, help advisors to at least ask the right questions of a student who wishes to pursue a full digital dissertation.

The Digital Dissertation: Archive?

Researching, planning and producing a dissertation is difficult enough, and adding a digital component increases the difficulty considerably since the author must not only make a conceptual contribution but must also reckon with the formal considerations of the text. In fact, these formal elements are vitally important since they are challenged anytime one breaks away from the traditional dissertation format. That said, these unconventional formats have few if any models: while born-digital dissertations are becoming more numerous, they are often inaccessible and typically archived in analogue apparatuses rather than in their native format.[2] For example, a dissertation created in the web-based multimedia-authoring platform Scalar, is now archived in my university's library as a vast series of static images (JPGs), which are mere screenshots of each 'page' of the dissertation. Obviously, this renders much of the work inaccessible—there can be no dynamic text, no audio, no moving images, no roll-over displays, no working links, nor any real sense of the linking structure. In short, this archiving actually works in direct opposition to the very form of the dissertation which, if done well, is key to its conceptual framework.

The lack of access to completed, natively digital dissertations presents difficulties for graduate students as well as their advisors who seek guidance in the construction and defense of these unconventional texts.

---

2 Archiving dissertations in their native format is preferable but technologically problematic given platform obsolescence and the use of emulators that can run old software is cost prohibitive. See Chapter 4 of this collection for an overview of Gossett and Potts' efforts in this area. As it stands, unless students commit to hosting their own domains and keeping their dissertation updated, the work will become inaccessible. ProQuest, the main archive for dissertations in the United States, after blocking my own dissertation as well as a few others I know of, is now endeavoring to archive media-rich work in a compatible environment. I remain hopeful but skeptical. For more on archival projects, again please see Chapter 4 of this book.

This includes my own digital dissertation, which is barely archived and was completed in a client-based program that is not browser based.[3] Indeed, my nine-month struggle to retain my doctorate when I refused to offer a print-based, image-cleared version acceptable to ProQuest's archival policies ended only when I was able to convince all parties that the key arguments of the work would not hold up in an analogue environment. Quite frankly, the pressure to create some sort of print version was intense, but I had the luxury of being able to hold out and, as such, felt I could not back down, if only to establish a precedent for others whose fate was more precarious than my own.

After a brief overview of my path to establish context, I focus here on a rubric that has proven useful in evaluating digital scholarship for more than a decade.[4] Its parameters have been reviewed and streamlined slightly over the years and as a result, it offers just enough structure to ensure academic rigor, but is flexible enough to allow for fresh thinking and invention. In discussing these evaluation strategies, I hope to help make this work legible to the institutional parties involved in the granting of doctorates, and specifically to those involved in the dissertation process itself.

Visual Literacy in the Digital Age: My Case

In August of 2005, I successfully defended a media-rich digital dissertation in the Department of English at the University of Wisconsin, Milwaukee, after having justified its natively digital format to my dissertation committee.[5] A few days later, I began a postdoctoral appointment at the

---

3   My dissertation was created in TK3, a software program written in Smalltalk, a coding language that is no longer widely used. See below for a discussion of this choice.

4   See Virginia Kuhn, 'The Components of Scholarly Multimedia', *Kairos: Journal of Rhetoric, Technology, and Pedagogy*, 12.3 (2008), https://www.academia.edu/859541/The_Components_of_Scholarly_Multimedia and Virginia Kuhn et al., 'Speaking with Students: Profiles in Digital Pedagogy', *Kairos: Journal of Rhetoric, Technology, and Pedagogy*, 14.2 (2010), https://cinema.usc.edu/images/iml/SpeakingWithStudents_Webtext1.pdf

5   My committee was supportive after some initial skepticism and I intentionally chose a Chair who was the least tech savvy, figuring that if I could convince her, I could convince anyone. My committee included: Alice Gillam as Chair, Gregory Jay, Vicki Callahan, Charles Schuster and Victor Vitanza. I also had invaluable support from Bob Stein and his staff at the Institute for the Future of the Book.

Institute for Multimedia Literacy (IML) in the School of Cinematic Arts at the University of Southern California (USC), where I remain, joining the faculty in 2007. In the intervening years, I have confronted the need for assessment and validation of digital work on a regular basis. Indeed, I joined the IML as it was transitioning from a grant-funded research unit into an academic division, the sixth in the USC School of Cinematic Arts. As faculty in a professional school, albeit a top-ranked one, I have frequently had to explain the merits of my own work as well as that of my students to the more traditional constituencies of the University.

I direct a multimedia honors program, an interdisciplinary undergraduate curriculum that culminates in a media-rich senior thesis project anchored in the student's major. It is the first academic program created and housed at the IML and I began overseeing it just as the first cohort became seniors, ready to create their thesis projects in 2007. Despite curricular scaffolding and institutional support, guiding the construction of these projects was no easy task given the variety of disciplines represented, which ranged from Aerospace Engineering to Classics, from Biology to Theatre, Physics to Journalism. A good rubric was key and fortunately, we had one. Its parameters were established during my early days at the IML in collaboration with another postdoc, with input from faculty and staff. Originally, there were four areas with three sub-categories in each. After many years of trying to update and hone the rubric, in 2015, I worked with a particularly lively and intelligent cohort of seniors to revise it: we removed repetition and shifted emphasis slightly to reflect cultural and technological shifts that had occurred in the years since the document's creation. There are now three broad areas—Conceptual Core, Research Component, Form + Content—with three features articulated within each. Although this rubric was originally created for undergraduate theses, it has been used widely for born-digital scholarship of all types.[6]

I discuss the parameters separately, giving a salient example of the actuation of each in a digital text. This conceptual and formal overview has been quite productive in the many production-based classes I teach as well as in the workshops I have done with faculty at several

---

6   Perhaps the best example of this rubric's adoption can be found in Cheryl E. Ball, 'Assessing Scholarly Multimedia: A Rhetorical Genre Studies Approach', *Technical Communication Quarterly*, 21.1 (2012), 61–77, https://doi.org/10.1080/10572252.201 2.626390

institutions in addition to my own. Obviously, it is more difficult to effect this overview on the printed page with only words and static images, but it is a useful exercise in translation that digital scholars will need to become practiced in, at least until these born-digital texts become more widespread and better understood: the more dynamic facets and the more subtle aspects of a digital text are difficult to describe and are better experienced, or at least witnessed during navigation.

I. CONCEPTUAL CORE

- The project's controlling idea must be apparent and be productively aligned with one or more multimedia genres.
- The project must approach the subject matter in a creative or innovative manner.
- The project's efficacy must be unencumbered by technical problems (which typically involves having a back-up plan).

Thesis driven prose is not the only, nor often the best option for digital texts so, in lieu of a thesis statement, a controlling idea is helpful to keep in mind, especially as one gets into the weeds of producing a large and complex text. A sense of a conceptual core keeps one anchored, especially when the myriad formal possibilities arise. Staying grounded in a controlling idea offers some constraints, while it also requires one to recognize and avoid formal elements which are rhetorically crude. These include functions like blink tags in HTML, gratuitous animation in PowerPoint, overuse of the zooming function in Prezi, and incoherent use of transitions (like the infamous *star wipe*) in video editing tools. Including these features may show some technical know-how, but they will not demonstrate rhetorical prowess. In other words, their presence would merely constitute 'bells and whistles', unless the point is to show the range of possibilities available for expression, in which case, the justification would be the pivotal aspect. Indeed, the ways in which the controlling idea is served by the container in which it is presented should be explicitly discussed in an FAQ, or instructions for access, or a 'how to read this text' section, in all digital texts. In fact, I occasionally still find myself explaining *how* to navigate digital work that was published years ago, much of which pushed back against the sort of spoon feeding (e.g., 'click here!') that characterized many of these webtexts early on.

In addition to reckoning with the native functionality of a digital platform, one must also account for design issues such as color, font type and 'page' or screen layout. For instance, in a workshop on digital scholarship, a graduate student produced a hot pink screen that functioned as the landing page of her digital text. The screaming pink, which many in the workshop saw as 'gaudy' was, as the author explained, meant to express a feminist scream, a sort of primal anguish at having been left out of so much history. This was an excellent rationale and I use it here as a way of highlighting the fact that the 'productive alignment' referred to in this area does not necessarily mean imitating a genre, rather it means an awareness of one, whether one retains its conventions or subverts them. In the case of my dissertation, I sensed that a book-based metaphor was important to maintain in order to give readers a sense of its coherency with conventional formats, seeing the work as an extension of the standard, and this was key to my larger argument about an emergent language of images (see Fig. 1).

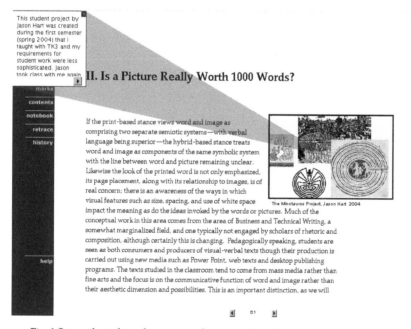

Fig. 1 Screenshot of my dissertation showing its book-based metaphor, as well as an 'annobeam', a function native to the platform which helps readers stay anchored in the main text, while giving the additional information the way a footnote does.

The efficacy issue mentioned in this area is obvious since a controlling idea must be legible and this means the container must be reliable or an alternative must be offered. We might think of this as the equivalent of grammar and typos in a word-based text; those formal elements, which work against the controlling idea, take us out of the conceptual space that the text should carve out and draw us into. On a trivial level, this was a problem for me because there was no spell check function in the software I used. In the larger scheme of things, however, this can mean offering video documentation of a particular function, such as projection mapping, which is notoriously tricky to use in situ. While theses and dissertations should offer new knowledge, some of which comes by formal means, the message is utterly lost if it cannot be accessed. Indeed, access issues (whether technological or human in nature) contribute to the multiple versions of a dissertation that many doctoral students, including several authors in this collection, have felt the need to create as a safety net.

II. RESEARCH COMPONENT

- The project must display evidence of substantive research and thoughtful engagement with its subject matter.
- The project must use a variety of credible sources, which are cited appropriately.
- The project must effectively engage with the primary issue(s) of the subject area into which it is intervening.

Obviously, all aspects of this area are important since conducting research is what distinguishes a doctorate from other terminal degrees. Still, the ways that research is expressed in digital texts can sometimes remain implicit, making its explication useful and sometimes necessary. The reference to 'substantive' research is tough to quantify and yet a comparison with the number of sources in a traditional dissertation could provide a solid roadmap. The 'credibility' of the source should not be an issue at this stage.[7] As for the 'variety' of sources, while this

---

7   Indeed, in my doctoral program, once students pass their preliminary or qualifying exams and become ABD, they receive a letter from the University stating that they are now considered to be a researcher.

verbiage was meant to remind undergraduates of the need to use some books rather than web-based sources exclusively, these days I often have to remind graduate students to use extratextual media sources in addition to books. If a digital dissertation is a valid text (and not simply a word-based text put online), then the author must also take other digital sources seriously even as these sources remain difficult to cite in conventional terms.[8] Indeed, it is important to cite not only word-based sources that are mainly conceptual in nature, but also media-rich ones that are invoked, or added directly to the digital dissertation. For instance, in my dissertation, I was pushing back against the convention of citing words but clearing images, often paying for this clearance but always asking permission, something I refused to do. I see this as a matter of free speech; if we fail to employ the language of images, then Hollywood effectively dictates who may speak and who is silenced. As such, using these media elements is crucial since the 'excess' of meaning encoded in non-textual media elements cannot be fully captured in words. Leaving them out, in my case, would have constituted incomplete scholarship.

This does not mean that anything goes with regard to the use of these media elements: we should exercise the same conventions as those we use for analogue media, and this means using only as much of the media element as is necessary to make a point, using the piece of media as an object of analysis rather than as decoration, and citing all sources. A discussion of sources and formats used is the sort of meta-level commentary that perhaps we ought to be having around *all* dissertations—if digital technologies allow the use of images and sound in addition to words, perhaps all dissertations should include a rationale for those registers that are deployed. In effect, this would find doctoral students justifying the use of only one register of meaning— the alphabetic. This is not to understate, in any way, the importance of verbal or 'natural' language in communication but it is to suggest that it is no longer the only game in town, as it were.

---

8   Typically, these dynamic texts are difficult to reference not only because they move but, more importantly, since it is difficult to direct a reader to a particular 'page' or screen. For instance, in Adobe Flash, once the standard for dynamic webtexts, there is only a single URL for the entire text. With HTML5 this is less problematic and on platforms such as Scalar, for instance, most of the Flash elements have been rewritten in HTML5, allowing multiple URLs.

III. FORM + CONTENT

- The project's structural or formal elements must serve the conceptual core.

- The project's design decisions must be deliberate, controlled and defensible.

- The project must achieve significant goals that could not be realized on paper.

In many ways, this area's focus on the relationship between form and content is the most straightforward one for a media-rich digital dissertation, since the format is the site of deviation and intervention. In terms of the reference to achieving goals that would be unrealized on paper, this was almost the default state of affairs in the late 1990s and early 2000s—simply putting a thesis in a form other than writing on a page made it innovative, and simply including the extra-textual registers of sound and moving image solidified this argument. And yet, this issue is far more complex than simply explaining the particular tool or platform used. There are very few platforms that allow the integration of media formats in a nuanced and sophisticated manner; nearly all digital authoring tools are either word-friendly or media-friendly but seldom both.

Creating tools is not really a viable option in the current state of late capitalism, at least in the US. While foundation money has been used to create tools like Sophie, an early version of which I used to create my dissertation, this sort of funding often dries up once the tool is created or the terms of the grant expire. Having worked at the forefront of tool creation for many years—in academia over the last two decades, as well as in the private sector from the mid-1980s to the early 1990s—I can attest to the fact that few accommodations are made for user testing and code debugging, never mind the increasingly short lifespan of any robust tool which requires nearly constant maintenance, all of which renders these tool creation projects problematic at best. The real issue, however, is the vast difference in scale of resources between such academic efforts and those of the tech giants today; stable software requires the multiple millions of dollars that tech companies are able to spend as well as the vast user base to whom they are able to feed updates. This means that awareness of the ideology around formal elements is key to any dissertation since nearly all scholarship is relayed through tools and platforms created with a free market, neoliberal ideology baked in.

Within the constraints of a particular platform however, there are also rhetorical approaches to its affordances, and the more explicit an author is about these choices, the better. An example of this comes by way of a webtext published in 2010 titled 'Speaking with Students: Profiles in Digital Pedagogy,' in *Kairos*. This particular webtext is quite germane to this discussion on many levels: it features overview videos of the students discussing their born-digital, media-rich thesis projects, as well as contextual information including the rubric under consideration here. The 'pages' of the webtext (which was created in the now obsolete program Adobe Flash), were carefully crafted to be semi-opaque, resulting in the presence of 'ghost' images—those of the other pages behind it (see Fig. 2). The journal's editors were initially concerned about this bleed-through, which they read as a mistake, and asked us to fix it. But we explained that this was done with great intentionality as a visual indication of the sort of 'thickness' of the more spatially-oriented texts that we felt were just on the horizon, via more accessible 3D modeling programs, Virtual Reality and Augmented Reality gear,

Fig. 2 Screenshot of a webtext published in 2010 in *Kairos* ('Speaking with Students'), in which the opacity has been carefully controlled to show the ghost of other pages in order to indicate the third dimension.

and consumer-grade depth cameras which can render 3D images from mobile phones. The journal editors were very supportive and simply asked us to add this information to the webtext, and we placed it at the end of the introduction.

*****

In retrospect, overseeing those inaugural thesis projects without a definitive template or archiving scheme was opportune, because it allowed experimentation with form when the stakes were not so high for either myself or the students. Indeed, when the first cohort of thesis students was graduating, I had met with university librarians in order to figure out how best to archive these projects. There was no viable plan and since this was undergraduate work, there was no mandate to 'publish' these in the University's library system. As a solution, we moved toward project documentation, and since the IML had enough resources at that time to create the five-minute documentation videos published in the webtext, we were free to experiment and explore. The rubric allowed us to speak to projects across a range of topics and disciplines, crafting these videos with some uniformity.

I have directed these multimodal, interdisciplinary undergraduate theses for more than a decade and have served on numerous doctoral committees in multiple disciplines, with dissertations that are either fully word-based (aka traditional) or are hybrid in nature and include a media object accompanied by a monograph that lends theoretical grounding and interpretation. These hybrid dissertations are quite similar to the sort of arts-based research that has thrived in Europe for many years; in this scenario, the critique or explanation or interpretation is done outside of the object, and the object is sometimes included as supplementary material. In other words, the written portion could ostensibly stand on its own, given an adequate description of the object of analysis. In the humanities, however, one of the most traditional academic areas, this seems like a far less acceptable solution.

My focus is on this particular rubric because I have employed it for many years and can attest to both its theoretical and practical use. That said, there have been some really promising assessment schemas for digital scholarship formulated and gathered by prominent scholars such as Todd Presner, Cheryl Ball and Anke Finger. These efforts are extremely important in terms of moving digital scholarship from the margins to the center and removing the remaining stigma around its authorship as well as its value as a scholarly object, rather than mere curiosities that seem compelling or interesting to note, but not engage with any depth. But perhaps more vitally, they are absolutely essential to helping move the humanities away from the fetishization of the single-authored (word-based) monograph as the only and most valuable form of artistic and critical scholarly work.

A Final Note

In closing, I want to briefly mention that while rubrics are helpful in terms of assessing a finished product, and of course they should be kept in mind at the planning stages of a dissertation, they are not as helpful when guiding revisions of a work in progress. Giving notes on digital texts is not only a conceptual difficulty, but it is also a logistical one when we cannot simply pick up a pen and add notes or even type digital comments into a word processing program. The conceptual difficulties will surely become less pronounced once more senior scholars are conversant with these texts, but in a practical sense, how do we handle giving notes on the extra-textual registers—video, audio, linking structures and the like—without either spending an inordinate amount of time explaining notes in words, or by conducting the review in real time with the student author present? Having responded to hundreds of video essays over the last fifteen years, I have spent an exorbitant amount of effort making detailed notes, marking time codes for each, and then trying to describe the sorts of complex media revisions I am suggesting using words alone, so I have a vested interest in finding better methods.[9]

The raw truth is that as digital scholarship becomes more sophisticated and more ubiquitous, so too will its editing and revision processes. The extremely time-consuming act of commenting on texts that work across the registers of word, images, sound and interactivity will certainly not be lessened, and will likely be far more involved. As a

---

9   I must give credit to Cheryl Ball, and her feedback on the first video-heavy collection I published in *Kairos* in 2008. Working with the fifteen pages of notes which she assembled to help in my revision of each video in the collection taught me how to give effective notes myself. Although it took me years of practice to become really good at giving these notes, it would have been impossible without that early model.

starting place for thinking about a robust revision process that does not prove overly onerous for either party, we might look to peer-reviewed digital journals and their submission process. I have served as a reviewer for a number of digital journals over the years, but I have experienced both sides of the review process with work published in *Kairos: Journal of Rhetoric, Technology, and Pedagogy*. Conceptually, the review notes I have been given on digital work from *Kairos* reviewers have been extremely influential for my own peer reviews, as well as for my notes on students' digital work. Logistically, I find promise in a plug-in developed for the *Journal for Artistic Research* (*JAR*) which allows a reviewer to comment directly into the text. This plugin is accompanied by an excellent review form that resists a stable set of criteria but, like *Kairos*, asks instead about the text on its own terms. Such questions include: *Which aspects of the submission are of interest/relevant and why?* and *Does the submission live up to its potential?* These questions allow for inventive texts while also recognizing they can almost always be improved upon. But perhaps more importantly, by linking student work to professional work we remember the ecosystem of institutions of higher education and remind ourselves of the value of our work with students.

Bibliography

Ball, Cheryl E., 'Assessing Scholarly Multimedia: A Rhetorical Genre Studies Approach', *Technical Communication Quarterly*, 21.1 (2012), 61–77, https://doi.org/10.1080/10572252.2012.626390

Editorial Board and the Review Process, *Kairos: Journal of Rhetoric, Technology, and Pedagogy*, http://kairos.technorhetoric.net/board.html

Finger, Anke, *Digital Scholarship Evaluation*, https://dsevaluation.com/

Journal of Artistic Research, 'Peer Review Form' (2019), https://jar-online.net/sites/default/files/2019-12/JAR_Peer-Review_Form%202019.pdf

Virginia Kuhn, 'The Components of Scholarly Multimedia', *Kairos: Journal of Rhetoric, Technology, and Pedagogy*, 12.3 (2008), https://www.academia.edu/859541/The_Components_of_Scholarly_Multimedia

Kuhn, Virginia, et al., 'Speaking with Students: Profiles in Digital Pedagogy', *Kairos: Journal of Rhetoric, Technology, and Pedagogy*, 14.2 (2010), https://cinema.usc.edu/images/iml/SpeakingWithStudents_Webtext1.pdf

Presner, Todd, 'How to Evaluate Digital Scholarship', *Digital Humanities Quarterly*, 1.4 (2012), http://journalofdigitalhumanities.org/1-4/how-to-evaluate-digital-scholarship-by-todd-presner/

4. #DigiDiss

A Project Exploring Digital Dissertation Policies, Practices and Archiving

Kathie Gossett and Liza Potts

The Digital Dissertation project, often referred to by the hashtag #DigiDiss, began with a research study in 2008 to better understand the needs of students, faculty and administrators composing and advising born-digital dissertations. In 2012, it expanded to designing and developing tools to advance their efforts and continues to study the emergence of this modality of academic discourse. In this chapter we present the exigence for the project, describe a workshop sponsored by the National Endowment for the Humanities (NEH) focused on gathering requirements for a digital dissertation repository, and briefly touch upon the latest phase of the project, a partnership with the Humanities Commons, as we endeavor to stitch these ideas together into a functioning process and a supportive, long-term network for scholars.

The Need for the #DigiDiss Project

The digital humanities (DH) are increasingly leading the research, discussion and dissemination of scholarship highlighting how computers and computer-enabled technologies transform traditional media and contribute to the production of new modes of expression. Institutions of higher education have responded by creating DH centers and doctoral-level programs in digital media and instructional

https://doi.org/10.11647/OBP.0239.04

technologies. Researchers in these fields are not simply concerned with studying and describing the phenomena; they seek to perfect the various techniques used to produce digital media, and subsequently use them to interrogate the usual modes of academic inquiry. Yet, despite a growing acceptance of digital media as a form of academic expression, the dissertation, even within DH fields, remains primarily print-based. This is not because doctoral students or committees are unwilling to consider born-digital projects—projects that are conceived and authored as works of digital media—rather, the reticence stems from the fact that there is no mechanism to adequately archive and publish such projects, a requirement at the majority of PhD granting institutions.

At the time of the project, ProQuest/UMI Corporation enjoyed a near monopoly in dissertation publishing in the United States through legal arrangements negotiated with doctoral-granting institutions. ProQuest was just beginning to pilot a system through which doctoral candidates could submit and publish their dissertations digitally, but it only allowed them to do so via the proprietary PDF format developed and maintained by the Adobe Corporation. Even as current PDF formats allow for the embedding of certain types of media (e.g., URLs, images and video) ProQuest's digital option continues to allow only for a print-based model of publishing focused on words. Since many works of digital media conceive of words as simply one of a number of modes that are integrated into complex visual, audio and interactive forms of digital performance, these requirements can impose considerable impediments and even misrepresentations, undermining the overall message of scholarly work. In effect, these requirements are obsolete. And since publication through ProQuest is often mandated by doctoral institutions as a condition for successful graduation, doctoral candidates often find themselves having to produce two versions of their dissertations: one representing their born-digital scholarship (e.g., interactive webtexts, software, apps, games, etc.) and another satisfying the need to deposit the dissertation into an archive (e.g., print-based PDF, etc.).

As an alternative to ProQuest, many institutions began installing and maintaining their own digital archive systems. The most common system in use at the time the #DigiDiss project started was Virginia

Tech's Electronic Thesis and Dissertation (ETD) system.[1] Much like ProQuest, the ETD system privileged print-based formats over multimedia or interactive formats. Although it was possible to deposit a majority of digital formats in an ETD system, file size and quantity restrictions meant that most born-digital projects had to be condensed into an archived file type (e.g., .zip or .dmg), requiring future readers to download and expand the project before accessing it (assuming the software is not out of date). In addition, the ETD system was a turn-key system;[2] that is, each university purchased, installed and maintained a unique instance of the system for their campus; therefore, unless the university decided to participate in one of catalogs maintained by the Networked Digital Library of Theses and Dissertations (NDLTD), there was (and still is) no central repository or search engine for ETD systems. The participants in the workshop during Stage 2 of the project explored the possibilities for building on the NDLTD framework to develop a national open-source and open-access archive as well as brainstormed ways to maintain the archived projects so that they remain accessible beyond current versions of software and coding languages—something neither ProQuest nor the ETD system do.

Project Stages

The core project team, Kathie Gossett, Liza Potts and Carrie Lamanna,[3] came from varied backgrounds in both industry application and academic research. Two of us encountered barriers for producing digital dissertations in our home institutions for various reasons (policies, time, access), and we determined to continue to look for ways to support scholarly research that results in born-digital dissertations and other digital scholarship across our disciplines.

---

1   Throughout this chapter, when the term ETD is used it refers to this original program developed and disseminated by Virginia Tech, which has since evolved into VTechWorks.

2   A 'turn-key system' is a program or system that is ready for immediate use as it comes 'out of the box'. It may support some customizations, but it is not designed or developed 'from scratch for different clients. Content management systems such as Canvas and Blackboard are turn-key systems.

3   Carrie Lamanna was the co-PI on the project with Kathie Gossett in 2008 and was a participant in the workshop described below. She left the project team after that stage.

This project began in earnest in 2008, and is ongoing with the following stages:

Stage 1: 2008–09—Study and field survey. Investigated the barriers graduate students and faculty advisors encountered when attempting to complete a digital dissertation. Conducted by Kathie Gossett and Carrie Lamanna.

Stage 2: 2012—Workshop sponsored by the National Endowment for the Humanities at Michigan State University. Gathered the first set of requirements and identified stake holders for a digital dissertation repository. Led and facilitated by Kathie Gossett and Liza Potts.

Stage 3: 2013–17—Thrashing and general weeping. Figuring out storage, networking, technology and other design issues. Work conducted by Kathie Gossett and Liza Potts.

Stage 4: 2018+—Working with Humanities Commons. Research and prototyping for implementation conducted by Kathie Gossett, Liza Potts and Kristen Mape.

Stage 1: Study and Field Survey

In 2008 and 2009, Kathie Gossett and Carrie Lamanna conducted a study into the status of digital dissertations in the field of Writing Studies. They collected survey and interview data from both faculty and graduate students—the majority of whom were drawn from the sub-discipline of Computers and Writing—regarding their experiences with born-digital dissertations. The findings, presented at both the Conference on College Composition and Communication and Computers and Writing in 2009, were surprising. Despite the early history of born-digital dissertations in the field (e.g., Christine Boese's 1998 dissertation)[4] and the well-known example of Virginia Kuhn's 2005 dissertation at the University of Wisconsin, Milwaukee,[5] the support for such types of dissertations

---

4   Christine Boese, 'The Ballad of The Internet Nutball: Chaining Rhetorical Visions from the Margins of the Margins to the Mainstream in the Xenaverse' (PhD dissertation, Rensselaer Polytechnic Institute, 1998), http://www.nutball.com/dissertation/

5   Virginia Kuhn, 'Ways of Composing: Visual Literacy in the Digital Age' (PhD dissertation, UW-Milwaukee, 2005).

was still very problematic. In fact, of the twenty-four graduate students interviewed for the project, all of whom identified themselves as planning to complete a born-digital dissertation in the initial survey, only two actually completed their project.

The study found four key obstacles to digital dissertations: 1) most graduate curricula, even in digital media-focused programs, did not include courses in digital authoring, thus requiring students interested in pursuing this type of scholarship to spend extra time (often years) learning the technologies they needed to complete their dissertation work; 2) a lack of institutional policies regarding born-digital dissertations; 3) the vast majority of the faculty had no experience evaluating digital work and did not feel qualified to do so; and 4) the inability to deposit or archive the digital work (often a requirement to complete a doctoral degree), since, at the time, the majority of venues for depositing dissertations did not accept born-digital dissertations.

At the time, we (Gossett and Potts) worked together in an English department at a medium-sized, east coast institution that had a graduate program in new media, where we were supervising students whose dissertation projects should have been either fully or partially born-digital. It quickly became apparent that the largest barrier to these projects at the institution was the requirement to deposit dissertations with ProQuest, which did not accept born-digital dissertations. (As noted above, ProQuest did accept 'mediated' PDFs, that is PDFs with embedded links or other media, but it did not accept born-digital projects such as websites, animations, videos, etc.) Given our industry backgrounds in user experience (UX) and software development, we decided we would build an open-source, open-access repository for dissertations.

Stage 2: Workshop with Stakeholders

In 2012 we, Gossett and Potts, received a level two start-up grant from the NEH's Office of Digital Humanities to hold a three-day workshop at Michigan State University. We gathered thirteen participants from across the United States—scholars, librarians and graduate students in DH, library and information sciences, and digital publishing—and asked them to help us identify the issues, opportunities and requirements for developing an open-source and open-access system into which born-digital dissertations (e.g., interactive webtexts, software, apps, games, etc.) could be deposited and maintained, and through which they could be accessed and cross-referenced.

The three-day workshop utilized UX methods to gather data about existing systems as well as identifying key users and stakeholders for the project and to begin identifying system requirements for a digital dissertation repository. Throughout the workshop participants cycled through group discussion, tool critiques and breakout sessions to articulate key issues, discuss limitations and possibilities for solutions, and created a first-cut needs assessment and conceptual design for a digital repository for born-digital dissertations.

Day One: Defining Key Concepts and Landscape Analysis

During the first day of the workshop, we introduced the project and defined key concepts. With the participants, we performed a landscape analysis to better understand how digital dissertations were being produced, supported and submitted across DH and humanities programs. This method is a process of analyzing the competition and identifying best practices so that designers can gain a better understanding of how a system should function. The workshop participants began this process by brainstorming a list of systems currently in use at universities to archive digital scholarship and/or dissertations, developing a list that included: Collex, Fedora Commons, RU Core, Digital Commons, DSpace, ETD, Content DM and GIT Hub. Subsequently, the participants examined characteristics of each system such as the ability to embargo/restrict access to the digital work for a specific period of time, the ability to perform a faceted search of the digital works, the depth of the metadata capabilities of each system, and whether or not the system(s) was open-source and/or open-access.

Based on the inventories collected during the landscape analysis, the workshop participants went on to compile preliminary requirements for a possible digital dissertation repository. These requirements included features that we thought the system should have (e.g., a federated search mechanism, responsive web and server design, metrics for tracking use of the system, etc.) as well as some of the challenges these features might pose (e.g., aligning institutional priorities with discipline-specific priorities, maintaining—not just archiving—digital artifacts, whether the system should follow a federated or single-source model, etc.). This led to a discussion on possible users/participants/stakeholders of these systems, which include department chairs, graduate deans, dissertation committees, graduate students (current as well as future), research assistants, librarians, university CIOs, provosts, IRB committees, publishers, DPLA, scholarly societies, research grant agencies, research participants/subjects and other databases (e.g., LexisNexis, ERIC, etc.).

Day Two: Developing ANT diagrams, Needs Assessment, and Personas

During day two of the workshop, we created rough versions of actor-network theory (ANT) diagrams,[6] conducted a preliminary needs analysis, and outlined personas based on our workshop participants' brainstorming, discovery, and discussion. First, we walked through the process of creating ANT diagrams (a design methodology developed by Liza Potts and based on actor-network theory). These diagrams help teams document all of the actors (people, places, organizations and technologies) that will be involved in the proposed system.[7] By visualizing these ecosystems, design teams can better understand the spaces in which a new technology will be deployed. Because the context in which digital dissertations are developed, defended and deposited are extremely complicated and often unclear, these diagrams were our first step towards better understanding the problem space from the perspective of our workshop participants (i.e., one set of project stakeholders). They proved to be an excellent brainstorming activity for our participants, as each worked to come up with a central figure that would work within the proposed system (e.g., the dissertator) and devise other actors who might support or even hinder their work (e.g., the dissertation chair/committee).

Next, we took these ANT diagrams and used them as a way of understanding the needs of the multitude of people and organizations

---

6   Liza Potts, 'Diagramming with Actor Network Theory: A New Method for Modeling Holistic Experience', *Proceedings of the IEEE International Professional Communication Conference* (2008), 1–6, https://doi.org/10.1109/IPCC.2008.4610231

7   Ibid.

participating in these spaces. Conducting a needs analysis means that we researched, discussed and documented the strengths, issues, concerns and weaknesses of all of the relevant actors in the system. Workshop participants took turns discussing the various needs, policy considerations and administrative constraints under which each proposed user would need to operate. One of the tools used to help workshop participants better understand user needs was *empathy mapping*, which assists designers both in gaining a deeper understanding of users as well as in identifying gaps in their understanding of users.[8]

Finally, we used the ANT diagrams and the needs analysis to help us decide which people and organizations required critical attention in order to launch any proposed solution. From this data we began to work on personas. Personas are applied in UX research to help design and development teams get a clear picture of who would employ a specific system and how it would be utilized. They tell the story of the central participants that any new technology or process would need to support. Although we knew we would eventually have to go back and refine these drafts, day two allowed us to gain valuable insights from our workshop participants and co-create this material in close collaboration.

Day Three: Identifying Next Steps

After debriefing the work of the previous two days, the third day's focus was on next steps. Participants brainstormed and made lists of potential future participants and advisory board members, as well as target grants, funding agencies and publication venues.

Workshop Findings and Yield

The workshop provided an excellent opportunity to bring together senior and junior scholars, graduate students and academic professionals to discuss the needs, issues and opportunities for archiving digital dissertations. While preparing for the workshop, we were very optimistic about the depth of scope for the workshop. During the workshop itself we quickly realized that the subject-matter experts were best situated

---

8   See David Gray, Sunni Brown and James Macanufo, *Gamestorming: A Playbook for Innovators, Rulebreakers, and Changemakers* (Boston: O'Reilly Media, 2010).

to discuss stakeholder needs, best practices and university procedures more so than design a system. We were able to shift into discussing the process of developing and implementing the system, rather than focusing on the user interface, database structure or information architecture of such a system. This kind of guided conversation led us to understand that there was a need not only for an archiving system, but that it should be a federated network of networks (i.e., a system installed and maintained at individual institutions networked together rather than one central repository installed and maintained at a single institution) that could catalog these dissertations.

After the workshop we, Gossett and Potts, spent several months analyzing and categorizing the research and data gathered during the workshop. Ultimately, we developed the four major findings below and reported them in the project white paper for the NEH.[9]

System Features and Best Practices. These practices included a federated search mechanism, responsive web and server design, and metrics for tracking use of the system.

Potential System Challenges. These challenges include aligning institutional priorities with discipline-specific priorities, maintaining—not just archiving—digital artifacts, and whether the system should follow a federated or centralized model.

Project Stakeholders. Workshop participants identified project stakeholders. The list of stakeholders included those internal to the university (e.g., provosts, department chairs, graduate students, etc.) and external to the university (e.g., governmental funding agencies, external systems, industry recruiters, etc.).

Project Partners. Workshop participants analyzed the market for digital dissertation systems and discovered some existing areas of opportunity for a repository and, thus, several potential partnerships with existing systems. In addition, the workshop group spent the majority of day

---

9   For further detail, including the landscape analysis data, ANT diagrams and personas described in this section, refer to our NEH white paper, 'Building an Open-Source Archive for Born-Digital Dissertation', *NEH White Paper* (2013), https://bit.ly/3lxxODr

three brainstorming a list of potential partners and strategic alliances for a digital dissertation repository project in the future.

Stage 3: Storage, Network, Technology and Design Concerns

Clearly, given the climate for digital dissertations and the technological shifts that were and are occurring, the technology, processes and policies had to catch up with the needs and desires of digital humanists. Faculty members were just beginning to appreciate the amount of work, time and effort that it would take to create a digital dissertation, especially if the dissertation was to remain a solo endeavor. Students lacked access to examples. Administrators and faculty wanted to ensure that students would be able to make their work legible to hiring committees while also displaying the training to produce both future multimedia and traditional scholarship.

In the meantime, we came to believe that a strong network that would live beyond the dissertation moment would potentially outweigh the need to house and archive the dissertation materials. It marked the moment the core team turned from designing a *discrete system* towards thinking through what a *networked system* that would link scholars and their digital scholarship would look like. We worked with partners in the Michigan State University library to brainstorm ideas about these kinds of networks, debating design ideas that would work for DH. We thought through how and why someone would use the system and designed multiple versions of wireframes, low fidelity drawings that depicted what each screen in the system would look like. We built prototypes and tested them. And we continued to iterate each time we became aware of technologies that would simplify processes or solve problems we identified.

In the end, we realized that simply creating a new system or piece of technology was not the answer to the problem, nor was creating a new network. Networks such as Mendeley, Academia.edu and others had begun to emerge as places for academics to store and share their work, as well as connecting it to the work of other scholars. These spaces were making the dissemination of scholarship open-source *and social*. We realized that we needed to wait and see what the academic community would do with these spaces.

In one sense this pause might mark the #DigiDiss project as a failure; we did not create the tool we set out to design. However, in the process of trying to create one specific tool, the research we did and the issues we uncovered revealed that the challenge of born-digital dissertations could not be solved by creating a new system. While some of the issues we identified required technology that had simply not been invented (and in fact some have yet to be invented at the time of this writing), we came to understand that the challenge of born-digital dissertations was more complex than that. Longstanding institutional attitudes and habits still remained. The issues encountered by students and faculty in the original study in 2008–09 were still problems a decade later. Many institutions had moved forward technologically by creating digital depositories for scholarship and dissertations, but some academic attitudes towards born-digital scholarship had not moved forward with it. At the same time, new academic social spaces were giving scholars and graduate students ways to disseminate and control their scholarship outside of traditional scholarly venues (e.g., pay-wall blocked journals and archives).

One of the guiding principles of UX design is to 'put human needs, capabilities, and ways of behavior first, then design to accommodate those needs, capabilities, and ways of behaving'.[10] Through the iterative process of ethnographic-style research and design work we did for this project during this stage we came to realize that building a digital dissertation repository would not solve the true needs of the majority of the stake holders we identified on the first day of our workshop in 2013 (e.g., department chairs, dissertation committees, graduate students, etc.). So, we opted to suspend working directly on the #DigiDiss project while continuing to track how the academy began to use the social media-based archiving systems, both those already in use and those that were emerging at the time.

Stage 4: Partnering for Network Stability and Sustainability

While we brainstormed, prototyped and considered implementation solutions, technologies advanced and the field progressed in its thinking about digital scholarship. New networks began to emerge

---

10   Don Norman, *The Design of Everyday Things* (New York: Basic Books, 2013), p. 8.

and gain currency. One of those networks was the Modern Language Association's (MLA) Humanities Commons (HC), which has since moved to Michigan State University.[11]

The HC is based on the Commons in a Box platform originally developed at City University of New York and the CUNY Graduate Center. It is a 'nonprofit network that enables humanities scholars and practitioners to create a professional profile, discuss common interests, develop new publications, and *share their work* [emphasis added]'.[12] It is an open-access, open-source, and non-profit space owned and governed by academics. In addition to the social media/sharing aspect of the system, it is built around the Commons Open Repository Exchange (CORE), which 'allows users to preserve their research and increase its reach by sharing it across disciplinary, institutional, and geographic boundaries'.[13]

By 2018 the HC had emerged as *the* space for humanities scholars to gather online and share their scholarship. More, the CORE system met or exceeded the requirements we had developed for the digital dissertation repository. While the HC was primarily targeted to scholars in humanities fields, researchers from across the disciplines were joining and depositing their work in the system. So, in the Spring of 2018, we partnered with Kristen Mapes at Michigan State University to pursue new implementation possibilities.

The Case for Humanities Commons

Potts began exploring the HC on the advice of Mapes, a DH specialist in the College of Arts and Letters at MSU. Recognizing the HC network and the archive as a powerful combination for a possible digital dissertation repository solution, we decided to proceed with an HC proof of concept and met with Kathleen Fitzpatrick, the Director of Digital Humanities at MSU and lead of the Humanities Commons. Over the course of several

---

11   'In November 2020, the fiscal responsibility and hosting of Humanities Commons moved to Michigan State University, where the network is developed and maintained by members of the MESH Research team' (Humanities Commons, 'About Humanities Commons', *Humanities Commons* (2016), https://hcommons. org/about-humanities-commons/).

12   Ibid.

13   Ibid.

weeks, the team created a project brief they could deliver to Fitzpatrick and the HC team. After their approval, the team proceeded with the first stage of the project which had shifted from an emphasis on archives to a focus on networking and linking scholars to their digital authorship as well as to each other.

The second stage of the project focused on the processes for using the HC. Working with research participants at an exemplar university, Mapes conducted stakeholder interviews and focus groups to learn more about their process for using the HC to create their school network and the HC repository for submitting, archiving and cataloging their dissertations for humanities, social science and STEM disciplines. Based on the data Mapes brought back, we worked with her to create interface prototypes. We tested these prototypes and made recommendations to the HC to implement those prototypes. We hope to continue our work with the HC stakeholders to create a content strategy aimed at dissertating students, their faculty advisors and university administrators who are interested in using HC for their network and repositories in the future.

Conclusion

The #DigiDiss project, which for us includes all stages of the project, is an example of the excitement and the perils of DH work. Begun in 2008 as a mixed-methods research study, we chose to focus on what we thought was the most 'solvable' of the findings of that project: the inability to deposit or archive the digital work. The turn of many humanities scholars to learning code and developing scholarly tools made it possible for us to determine to build the archive that was missing: a digital dissertation repository. Given both of our backgrounds in software development and UX in industry, we felt that this was a project we could take on and guide to fruition with the help of some of those scholars.

It was an exciting moment. We were invited to speak on panels at multiple conferences and were invited to give the keynote address at the 'Research in the Digital Age Symposium' at Trinity College Dublin, Ireland in 2015. Digital dissertations seemed to be emerging as an acceptable form of the dissertation in the United States and internationally. The #digidiss twitter feed was active and trending within academic communities. We connected with graduate students building digital scholarly editions, creating comics, building tools, making documentary films and writing and recording hip-hop albums as their dissertation projects. However, as we began the work of gathering requirements and designing the tool we envisioned, the perils of DH work began to emerge.

As we discussed above, we realized two things: first, some of the technologies we would need to make the system sustainable and successful were barely on the cusp of being developed; second, building the tool would not actually solve the larger problem. The institutional policies and attitudes toward digital scholarship at both the graduate and faculty levels were (and are) complex. Attitudes of faculty as well as the institutional policies that govern them and graduate dissertation projects are still evolving. Additionally, much of the work (and life) of digital researchers and scholars moved to networked (and social) spaces. While publication in peer-reviewed books and journals is still the accepted norm, many digital scholars have also chosen to share their work across open-source, open-access systems like the Commons networks. While we were focused on building the perfect repository for born-digital dissertations, networks like the HC built systems that supported the archiving and dissemination of digital research and scholarship of all types (i.e., born-digital and print-made-digital—such as PDFs). So, ultimately, we realized we didn't need another tool, we just needed to work with the HC to develop tools within their network for those we mentioned above: dissertating students, their faculty advisors and university administrators who are interested in using HC for their network and repositories in the future. Additionally, we continue to advocate for and encourage the development of born-digital dissertations at our institutions and in the academic societies in which we participate. The work continues, as do the born-digital dissertations.

Bibliography

Boese, Christine, 'The Ballad of The Internet Nutball: Chaining Rhetorical Visions from the Margins of the Margins to the Mainstream in the Xenaverse' (PhD dissertation, Rensselaer Polytechnic Institute, 1998), http://www.nutball. com/dissertation/

Gray, David, Sunni Brown and James Macanufo, *Gamestorming: A Playbook for Innovators, Rulebreakers, and Changemakers* (Boston: O'Reilly Media, 2010).

Humanities Commons, 'About Humanities Commons', *Humanities Commons* (2020), https://hcommons.org/about-humanities-commons/

Kuhn, Virginia, 'Embrace and Ambivalence', *Academe*, 99.1 (2013), 8–13, https://eric.ed.gov/?id=EJ1004358

Kuhn, Virginia, 'Ways of Composing: Visual Literacy in the Digital Age' (PhD dissertation, UW-Milwaukee, 2005).

Networked Digital Library of Theses and Dissertations, http://www.ndltd.org/

Norman, Don, *The Design of Everyday Things* (New York: Basic Books, 2013).

Potts, Liza, 'Diagramming with Actor Network Theory: A New Method for Modeling Holistic Experience', *Proceedings of the IEEE International Professional Communication Conference* (2008), 1–6, https://doi.org/10.1109/IPCC.2008.4610231

Potts, Liza, *Social Media in Disaster Response: How Experience Architects Can Build for Participation* (New York: Routledge, 2014), https://doi.org/10.4324/9780203366905

Potts, Liza and Katherine Gossett, 'Building an Open-Source Archive for Born-Digital Dissertation', *NEH White Paper* (2013), https://bit.ly/3lxxODr

Salvo, Michael and Liza Potts, eds, *Rhetoric and Experience Architecture* (Anderson, SC: Parlor Press, 2017).

VtechWorks, https://vtechworks.lib.vt.edu/

5. The Gutenberg Galaxy will be Pixelated or How to Think of Digital Scholarship as The Present

An Advisor's Perspective

Anke Finger

A PhD in the humanities traditionally requires a book-length study of original scholarship, aka a dissertation. As a matter of initiation into the world of academics, I wrote one, too, typing away at my Toshiba laptop, equipped with a feeble 120 megabyte hard drive. When I was not staring at the rhythmic heartbeat of the blue cursor on the screen that, I so hoped, would send sparks of life to my writing, I jotted down ideas on paper flash cards, to be used (or not) later. Digital text and ASCII code, that was the extent of the multimodal versatilities at my fingertips in the early and mid-1990s. Hyperlinks, images, graphs, video, audio, animation—common digital features in today's world—were absent. At the time, the book concept was easily transferable to early personal computing since the laptop, no matter how revolutionary its technical capacities at the time, produced text for which one simply did not have to use whiteout anymore. Paper saved, and typo nightmares and grammatical errors avoided with a simple click.

The book as a medium and as an artifact has changed significantly since then. According to Matthew Fuller, 'Nobody Knows What a Book Is Anymore', and he suggests that we consider the 'book as diagram':

> As we see books entangling with computational structures and entities we can perhaps see them undergoing a further transition: incunabula,

 https://doi.org/10.11647/OBP.0239.05

codex, book, stack, queue, heap. [...] The book is an essentially shifting, capacious form—there is not one aspect of its characteristics concerning binding, titling, authorship, typesetting, pagination, orthography, and so on, that has not been exceeded, gone beyond or done without in various and numerous cases.[1]

Few would dispute that, while the publishing industry is doing just fine producing print books, a plethora of digital book forms have emerged over the last twenty and more years, speaking to the enticingly experimental potential of what used to be called 'new media', but also to democratizing authorships and readerships beyond national, linguistic, economic and media limitations. What about academic books, however? The present in academia is not so innovative or manifold as of yet. Paul Spence lists a number of reasons for both resistance to and difficulties of producing digital book forms for scholarship. Among them he emphasizes the 'many challenges of technical sustainability and preservation, education and training, not to mention effective business models and integration into the wider fabric of scholarly communication'; a lack of understanding of 'the "digital book" (or its alternatives) as intellectual systems'; the meager number of 'studies regarding how digital publication actually facilitates or encourages new forms of knowledge production'; and a two-tiered and even oppositional relationship between print and digital forms.[2] He concludes that we have yet to figure out the 'many opportunities in fully integrating complex scholarly argument into a potentially more connective, participatory and visually expressive medium'.[3] If Spence dampens Fuller's perception of a rich and colorful landscape of book forms in the digital realm, Robert B. Townsend's 'Are Historians Still Ambivalent about Getting Published Online?' on the History News Network crushes anyone's enthusiasm about forging ahead for the future of academic digital scholarship and publishing. Based on a 2015 national survey conducted in history departments with and without PhD programs, he counted almost 80% of respondents who never published online because of the 'lack

---

1   Matthew Fuller, 'Nobody Knows What a Book Is Anymore', *Urbanomic* (2017), https://www.urbanomic.com/document/nobody-knows-book/

2   Paul Spence, 'The Academic Book and Its Digital Dilemmas', *Convergence*, 24.5 (2018), 458–76 (at 462–63, 466, 467, 471), https://doi.org/10.1177/1354856518772029

3   Ibid., 473.

of scholarly prestige'.[4] Over 90% confirmed that a print book is key for tenure. Notably, as Townsend points out early in his report, 'this ambivalence [about online publishing] appeared to arise from two principal sources—personal doubts about the value of this form of work, and a larger sense that there is little professional appreciation or credit for this form of work'.[5] I should note here that Townsend does not define 'online publishing' further, leaving the genres comprised by 'online publishing' wide open.

If historians were to resist, for example, blogging, web page design or hybrid outlets promoted by first-tier academic presses, how is the dissertation, as the precursor to an academic book (presumably with which tenure will be secured), to arrive at the digital stage? Why would anyone be reckless enough to put effort into the multitude of skills and hours needed to collaborate on and produce a dissertation in multimodal format? Why invest in so much technical knowledge and innovative energy when it is valued so little by those evaluating the work for one's future scholarly potential? As Virginia Kuhn succinctly put it in her article concerning the digital dissertation, 'the academy's resistance to the digital remains. [...] and tenure review boards have consistently shown themselves to be unprepared to reward or even credit junior faculty who produce digital scholarship'.[6] Kuhn here refers to her own 2005 dissertation, and, arguably, a considerable amount of time has passed since then, by digital measures. However, while hybrid or born-digital dissertations have appeared within the realm of possible humanities and art scholarship at many institutions, not much has changed in these years regarding evaluative measures. Most review boards continue to rely on the scholar's own explication of her or his work, and on a doctoral advisor's translational acumen, navigating traditional and multimodal approaches to scholarly communication. At issue are the variety of digital scholarship genres, formats or cultural techniques and collaborative work (which is standard in digital scholarship) that pose the most significant challenges for evaluating committees or units. These

---

4 Robert B. Townsend, 'Are Historians Still Ambivalent about Getting Published Online?', *History News Network* (2018), https://historynewsnetwork.org/article/168871

5 Ibid.

6 Virginia Kuhn, 'Embrace and Ambivalence', *Academe*, 99.1 (2013), n.p., https://eric.ed.gov/?id=EJ1004358

genres or cultural techniques are often unfamiliar, and collaborative authorship may remain a quantitative exercise in who did what and how much. The evaluation debates, however, also require a return to the most central of questions, namely: what is scholarship? What shapes and forms does it take now and in future decades? And who are its audiences? While many PhD advisors are digital-scholarship-positive or -curious, they may lack the training to guide the graduate student with expertise and themselves require assistance from numerous university networks. The graduate student, in turn, must learn new digital tools and methods, collaborate, write grants, and, importantly, become an adept communicator of one's own digital scholarship. They are obliged to explain their process, contrary to the traditional scholar whose methods and approaches are tacitly beyond reproach. These are time consuming and highly disruptive activities in addition to seeking employment within or beyond the academy or simply going about one's everyday teaching and research obligations. How can the academy provide a more supportive environment whereby the budding digital scholar is not also required to repeatedly defend and explain her or his process, methods, and tools?

In the following, I offer my perspectives as a PhD advisor and as the inaugural director of Digital Humanities and Media Studies at the University of Connecticut's (UConn) Humanities Institute[7] to suggest steps towards incremental change at the dissertation stage. For even at universities with limited tech support and no detailed guidelines on how to evaluate digital scholarship at any stage of academic research, such as the University of Connecticut, graduate students must receive access to scholarly inquiry and research innovation beyond print, beginning work with digital tools early in their graduate career, and move towards what Jeffrey Schnapp has called knowledge design.[8] Indeed, digital dissertations have been around for decades. More often than not, they must have been the product of a maverick or adventuring spirit who had the goodwill and generosity of an advisor ready to embrace their student's vision. Or the dissertation project was supported by an existing unit—a DH center or a digital lab—that provided the conceptual,

---

7   See https://dhmediastudies.uconn.edu/
8   Jeffrey T. Schnapp, *Knowledge Design* (Hannover: VolkswagenStiftung, 2014), http://jeffreyschnapp.com/wp-content/uploads/2011/06/HH_lectures_Schnapp_01.pdf

collaborative and tech support needed when fellow graduate students and, specifically, faculty advisors remained untrained in guiding the project and/or unsupportive of its epistemological endeavors. Smiljana Antonijevic, in her seminal study of that 'tribe' called digital humanists, repeatedly notes how often DH practitioners are self-trained and self-motivated, stoically weathering misunderstanding, dismissal or even ridicule of their work with digital media and computerization. At issue are cultural dissonances: peers and advisors maintain long-held values and practices in academia, with some unprepared or unwilling to adjust to means of communication and scholarly inquiry that move beyond print. Importantly, Antonijevic emphasizes, 'In discussing how to change this mindset [humanists' insular attitude toward the purpose of their work] my respondents commented further that these attitudes reflect economic circumstances and the overarching academic structure of tenure and career advancement in the humanities'.[9] Learning new digital tools and methods, collaboration, grant writing and, importantly, becoming an adept explicator of one's own digital approaches, are demanding activities in addition to what is assumed to be the focus of any ABD (all but dissertated) graduate student: researching and writing the dissertation such that a wad of paper, topped with a neat title page, will find its way to the graduate school for official approval towards the PhD degree.

Digital scholarship, if understood not only as working with digital methodologies and tools, but also as communicating and publishing beyond print media, presents the traditionally trained humanist with further challenges: how is the humanities scholar to navigate the plethora of media and media affordances? What about the variety of literacies required to read and produce such scholarship? How to negotiate the possible semiotic playing fields? Up for debate are not only local structures for digital scholarship (workshops, capable and supportive faculty and librarians, tools, equipment and archivists), but also continuing misconceptions or differing ideas about what constitutes digital scholarship in humanities and art departments, and, importantly, the necessity to dialog about what presents as an intercultural glitch

---

9   Smiljana Antonijevic, *Amongst Digital Humanists. An Ethnographic Study of Digital Knowledge Production* (London: Palgrave/Macmillan, 2015), p. 126, https://doi.org/10.1057/9781137484185

between two increasingly disconnected groups: those who 'do' digital scholarship and those who do not 'do' digital scholarship. Ultimately, I suggest, digital scholarship methodologies and practices continue to demand additional communication skills to translate between digital and analog epistemologies in humanities and art research. At minimum, it behooves advisors and faculty to equip graduate students with those skills so that they can advocate for themselves and their research; ideally, faculty and advisors would recognize at the local level that digital scholarship is very much The Present and adjust and update curricula and PhD programs accordingly.

Hybrid Dissertations

In the following I share a few humble first steps any instructor and advisor can implement into their graduate seminars or dissertation work to discover and explore approaches towards digital scholarship with their students and advisees. Over the years of advising graduate students towards authoring a dissertation, I began to integrate multimodal forms of expression and technical tools early in the graduate curriculum. First and foremost is the decidedly non-digital exploration of non-linear thinking. Different disciplines, including design and psychology, have established specific corpora of scholarship to explore this cognitive approach; in my case I employ the semiotics of multimedia or multimodality since my dissertation, and eventually book, took on the topic of the total artwork, requiring me to understand different codes and modes of communication in converged form. Nonetheless, a theoretical understanding of interart processes or word and images studies, for example, does not necessarily help with learning a certain middleware, as Johanna Drucker and Patrick Svensson explain,[10] nor does it teach one approaches beyond qualitative hermeneutics that take advantage of the computational, quantitative power of computer technologies. Or, more confusingly, how to design a product that employs the possibilities of non-linear, multi-layered and multimedia communication and design forms. A digital dissertation on the total

---

10   Johanna Drucker and Patrick B. O. Svensson, 'The Why and How of Middleware', *Digital Humanities Quarterly*, 10.2 (2016), http://www.digitalhumanities.org/dhq/vol/10/2/000248/000248.html

artwork in modernism using today's technical means would present as a carefully networked, intricately designed composition consisting of mp4 and mp3 files, enhanced by JPGs and text-mining graphs, and ngram-based data on the use of the term while urging the reader to cruise through the work using a variety of platforms. But how does one begin to think in this dimension?

Every graduate course I teach includes at least one media project feature. This media project is completely open, the only guidelines I offer are a) it must NOT be a linear text in print and b) the project content should be a first exploration of a possible topic for the final seminar project. Accompanying the project itself are 4–5 pages or about 1000 words of process writing, laying out explicitly how the author came up with the idea for the project, chose the medium/media used and why, and reflecting on mistakes and challenges along the way. I describe this process writing as a blueprint, should the author wish to produce the same project again, so that she or he can retrace these first, exploratory steps, both practically and theoretically. In my digital humanities seminars and for the DHMS Graduate Certificate I will describe below, I apply a more sophisticated model, derived from Shannon Mattern's foundational piece 'Evaluating Multimodal Work, Revisited'.[11] For students who have never embarked on multimodal scholarship (and considering the conundrum of multiple literacies), however, and who require assistance with stepping over a digital tech threshold for the purpose of producing scholarship, the intricacies of criteria laid out by Mattern's guidelines are far too complex. In that case, I offer Alan Liu's treasure trove of a toy chest with which students are emboldened to experiment, focusing on one or two tools of interest.[12] It encourages them to create a vast variety of works, from visual interpretations of texts to video to audio to games to installations. Some of them are completely new to the medium they produce: they have to familiarize themselves with the technical skills necessary to reach an audience (sound has to be audible, images have to be clear and used fairly, for example); they do the research to justify which (editing) tool they used; they are required

---

11   Shannon Christine Mattern, 'Evaluating Multimodal Work, Revisited', *Journal of Digital Humanities*, 1.4 (2012), http://journalofdigitalhumanities.org/1-4/evaluating-multimodal-work-revisited-by-shannon-mattern/

12   See http://dhresourcesforprojectbuilding.pbworks.com/w/page/69244319/Digital%20Humanities%20Tools

to communicate design decisions for a medium like a website, including color coding and wireframing; and they are asked to reflect on how this new medium helped them approach the topic or text at the base of their media project anew. Invariably, these projects become mini-independent studies. I help with technical issues, reframing questions, refer students to get assistance elsewhere or push them out of their intellectual comfort zones when there is fear of failure or mere frustration with the assignment. They all overcome the fear or concerns eventually since the assessment is not punitive: they get an A for this element of the seminar as part of their participation, provided they have an initial product they have reasoned though and applied a creative process that moved the epistemological bar to a next level. What is different is the critique: the process writing allows for deeper reflection on the making of, and especially the why, and the entire group critiques the final product such that the experimental nature of the assignment is embraced, not whether the video or audio is technically flawless or the topic itself is well-presented. Several students further refine the project, using the media affordances so effectively that the student could apply with it to one of their first conferences. Some, in my independent studies, for example, have produced an impressive corpus of data, complete with a thoroughly designed research approach, but need extra encouragement to present their work at meetings as it is considered 'unfinished'. It is this first adventure with digital scholarship that counts, it is the first application of digital tools that applies non-linear thinking and creativity, and it is the first exploration of nontraditional hermeneutics that—as they all avow—provides an entirely new perspective on the topic or text they chose to 'translate' in the first place.

I urge all dissertating students to apply this creativity as epistemology in their dissertations as well. Should anyone wish to write a born-digital dissertation, I am all for it. So far, most choose to stay either within the traditional parameters or they pick a chapter that becomes a digital humanities project, either accompanying the larger text of the dissertation or figuring as an integral part of the larger argument. The biggest challenge, I have found, is not the acquisition of new skills in the digital realm; students can build their own support system, and within our department, they have offered each other training on platforms or tools like Scalar or WordPress or Omeka or software languages. It is joining a community of practitioners, a new culture group, that speaks a different language and subscribes to and develops entirely new approaches to what we call literary or cultural studies.

Interculturally speaking, those who 'do' digital humanities and, by extension, digital scholarship move in a different communicative world that prevents uninitiated grad students just as much from approaching or being able to evaluate digital scholarship as it does established faculty. In May 2018, I was invited to speak about digital scholarship at an Association of Departments of Foreign Languages (ADFL) seminar for department heads in the languages, and, in preparation for my talk, I asked some Tenure and Promotion Review (PTR) committee members in my own department what they thought of digital scholarship. The responses were mainly positive, most were all for encouraging it; however, without fail, everyone was at a loss as to how to evaluate it. Not only did I come across readily admitted gaps in technical knowledge, there was also confusion as to how to 'read' a dissertation or book that clearly did not meet traditional parameters of peer review or metrics conventional in humanities and art scholarship.

Undoubtedly, the lingo can be daunting: in Johanna Drucker's reflection on 'Why Distant Reading Isn't', terms such as 'tokenization', 'probabilistic inference techniques', 'grayscale value', and 'ASCII string' will likely make most of my colleagues wonder how such terms figure in any part of their work, even if they are familiar with the practices of distant reading and data mining.[13] And a part of me does not want to bother them, for who am I to disturb the experts in the fields they have come to navigate superbly and for which I admire them as colleagues and fellow intellectuals. But can I? Do not we, as advisors, have an obligation to learn this language and culture of digital scholarship such that we can at least help guide those students who wish to move the profession forward with the tools that the twenty-first century provides them? Should we, as advisors, not at least foster digital dissertations as explorations into a different communicative world—especially in language and culture departments—such that our PhD students take full advantage of the intellectual and technical tools at hand to create for themselves novel academic and non-academic career paths? Who

---

13 Johanna Drucker, 'Why Distant Reading Isn't', *PMLA*, 132.3 (2017), 628–35, https://doi.org/10.1632/pmla.2017.132.3.628

are we, as advisors, to close ourselves off from a fully digitized universe in which learning and thinking and communicating has long embraced multimodal forms?[14] We need to develop reference materials and introductory guidelines for dissertation committees, such as the FICUS heuristic presented in the next chapter, as well as PTR committees that are far more detailed than the helpful, but locally and practically too nebulous guidelines laudably provided by the MLA (Modern Language Association), AHA (American Historical Association) or CAA (College Art Association). We need to equip advisors and faculty evaluators with insight into the language and culture of digital scholarship in practice and into its intellectual value. We need more reference anthologies like *Literary Studies in the Digital Age*,[15] or continued updates to foundational criteria, such as the 2012 list provided by Todd Presner,[16] where advisors and faculty evaluators gain access to information required for their work with graduate students and junior faculty. And we need administrators, at the very least department heads and staff at the graduate school, to fund and create repositories of such materials and sample works so that each institution can build case study histories that speak to the local evaluative culture and to the distinct disciplines within it.

The Making of Flusser 2.0—The Long Game

Obviously, as an advisor and a faculty member at a research institution one is to keep up within one's field(s). This may include learning a new language for a research project, applying a new sub-field or, in my case, making sure one has a nascent understanding of this area called digital humanities and digital scholarship. Trained in comparative literature and reorienting myself toward media studies, when I started out with

---

14   On the term multimodal, see Virginia Kuhn, 'Multimodal', in *Digital Pedagogy in the Humanities: Concepts, Models, and Experiments*, ed. by Rebecca Frost Davis, Matthew Gold, Katherine D. Harris and Jentery Sayers (New York: Modern Language Association, n.d.), https://digitalpedagogy.mla.hcommons.org/keywords/multimodal/

15   Price, Kenneth M., and Ray Siemens, eds (2013-present), *Literary Studies in the Digital Age. An Evolving Anthology* (New York: Modern Language Association), https://dlsanthology.mla.hcommons.org/

16   Todd Presner, 'How to Evaluate Digital Scholarship', *Journal of Digital Humanities*, 1.4 (2012), http://journalofdigitalhumanities.org/1-4/how-to-evaluate-digital-scholarship-by-todd-presner/

building rudimentary websites for scholarship and teaching, I drew inspiration from my own research to translate from print to digital as well. While I re-interpreted the 'correspondences between the arts' as a model for interarts communication and twenty-first-century cross-media relationships, my focus on the theory and analysis of art and media convergences eventually shifted to the making of multimodal scholarship (starting out with launching and co-editing an online journal, *Flusser Studies*, for ten years). The project presented here, *ReMEDIAting Flusser*, merges media studies—by focusing on the media philosopher Vilém Flusser—with digital humanities by building a multimodal e-book using Scalar, entitled 'Flusser 2.0: Remediating Images, Reimagining Text'. The project is collaborative, with three main contributors, a PhD student, an undergraduate and myself as co-constructors.

Building and collaborating are themselves considerable, nontraditional academic elements of scholarly work I had myself vastly underestimated. The 'invisible labor' behind such scholarship is significant, involving a creative and non-linear process that is recursively evolving, interactive, and multilingual. The Flusser project is conceived as open-ended and starts out with an introductory video, available on Vimeo.[17] This first element required learning how to write script, record technically adequate audio, acquire basics of Adobe Premiere Pro, and, with the help of a media agency, design meaningful and provocative correspondences between visual and textual codes that point to Flusser's philosophy of the technical image. The video itself has garnered 12,600 views, a whopping success considering metrics in the humanities.

The second stage consisted of editing and designing contributions from a variety of international Flusser scholars to structure and build the e-book in Scalar. The goal is to interconnect these contributions using a variety of media and to 'translate' core aspects of Flusser's philosophy into digital forms such as hypertext, visuals, video and audio. The final multimodal and multi-lingual product (Flusser wrote in four different languages) will consist of an interactive visualization of Flusser's main ideas, moving well beyond what he long ago identified as the end of the linear and alphanumeric code.

---

17   See 'ReMEDIAting Flusser', 5:19, posted online by Anke Finger (2016), *vimeo*, https://vimeo.com/156304605

The epistemological re-wiring necessary when engaging with multimodal critical making as scholarship is profound, both for myself and the now graduated PhD student I co-edit with, Britta Meredith. The various technical challenges inherent to Scalar, once one dives beyond the surface features, remain thorny. I received an internal grant to work with an undergraduate student, Katherine Riedling, on the coding, and she, too, grappled with D3 and the semantic web structures Scalar supports and that, we thought, so perfectly reflected both Flusser's thought processes and the network of his entire oeuvre. Britta and I, in turn, grappled with redefining the responsibilities of editing a digital scholarship project as well as time issues. We were a team of two for content and a team of three for tech elements: how did we want to divide the labor of editing content, including reviewing submissions, editing style, communicating with and encouraging (delayed) authors, writing the introduction and composing our own contributions? How deeply did we want to engage with the technical and structural possibilities of a platform like Scalar, including its annotation and interactive features? A print dissertation and book medium envisions an unknown reader, making integrated communication with an audience impossible. Given Flusser's philosophy of dialog, however, we also wanted to include a response or annotation feature, encouraging readers of the Scalar project to respond to us with their thoughts and ideas. We are not there yet. The coding is done, and Britta and I have managed to collect and curate the site such that all contributions are ready to be edited into their multimodal form (including links to video, audio, images and more). We have presented the project at different conferences, but time continues to elude us both to finish not 'just' the editing of the content, but also the designing of the Scalar product.

In fact, 'doneness', 'finitude', 'completion', within digital scholarship, also become fuzzy concepts, given the emphasis on process and collaboration. When we presented the most current version at a conference seminar focusing on digital humanities projects in progress, no one was more surprised than us when we were met with enthusiasm and a repeated refrain of 'this looks so done, you are so close'. Really? We did not think so, but perhaps we had also lost perspective? A project like 'Flusser 2.0', and, by extension, a digital dissertation, can quickly turn into the black hole that sucks up all energy and resources, not unlike the traditional dissertation. Only that we sweated over pixelation and D3 on top of citations, scholarly dialog and innovation. The proverbial 'the good dissertation is a done dissertation',[18] a phrase I have often repeated myself, inspired by my own dissertation advisor, becomes a hollowed phrase if you need to rely on other collaborators, an intercultural and interpersonal enterprise all its own, and your skill sets need constant updating and practice. Working within a program like Scalar requires engagement with intricacies of design and coding that are entirely absent from print publications. In short: we are not done yet. But at least Britta's contribution, a part of her dissertation, is composed in its full multimodal form, and it is by far the most 'done' part of the project.

Inaugurating DHMS

My experience with 'Flusser 2.0' and other projects I pursued over the last ten years influenced how I conceived of the new initiative of Digital Humanities and Media Studies (DHMS), launched in 2016. I became the brand new assistant director of the Humanities Institute with a $4,000 budget, in charge of what I named DHMS and fully responsible for development and programming. How did I want to entice graduate students and faculty to participate such that they saw digital scholarship as an endeavor worth pursuing in the humanities and arts? How to bring Digital Humanities and digital scholarship to UConn such that it becomes a viable, recognized and rewarded field of inquiry? Dan Cohen, in a summary blog post that is based on his 2017 talk at Brown University on 'Institutionalizing Digital Scholarship', identified three steps for sustainable DH initiatives: routinize, normalize and depersonalize.[19] In my case, finding space and support within a fully institutionalized unit such as the Humanities Institute, thanks to the director, Michael P. Lynch, was a major step towards visibility, since I was able to work from within the routines of the institute itself. This first step of securing a

---

18 Verena Kick, '"A Good Dissertation Is a Done Dissertation" — and Nothing Else Matters?', *HASTAC* (March 6, 2017), https://www.hastac.org/blogs/vkick/2017/03/06/good-dissertation-done-dissertation-and-nothing-else-matters

19 Cohen, Dan, 'Institutionalizing Digital Scholarship (or Anything Else New in a Large Organization)', *Dan Cohen* (November 29, 2017), https://dancohen.org/2017/11/29/institutionalizing-digital-scholarship-or-anything-else-new-in-a-large-organization/

recognized locale on campus appears to be particularly important since several colleagues from other institutions who consulted me wondered where to start looking for DH or digital scholarship support—which, in many institutions, means the library, or how to gather a community of interested faculty and graduate students, especially if the sheer size of the institution makes reaching beyond units difficult. An already established cohort of humanists was helpful and facilitated attracting an audience for talks and workshops. But how to routinize a practice of scholarship that was mostly unknown, sometimes mysterious or seemingly experimental? I created a multi-layered approach, focusing on building a network with regular meet-and-greets; organized regular roundtables (Fall) and talks (Spring) with well-known scholars in the field such as Kathleen Fitzpatrick, Cheryl Ball (both contributors to this collection) or Alan Liu;[20] collaborated with digital librarians to coordinate workshops and tech support; offered both resources and sample projects on the DHMS website;[21] and, most important, established a DHMS graduate certificate[22] that could be integrated with an MA or PhD program in the humanities and social sciences such that digital dissertations and scholarship would be supported.

Graduate students, once the certificate had met with approval from all necessary committees, regularly inquired about the course of study, with some unsure whether or not they would be able to squeeze more courses into their curriculum. Graduate students from different disciplines met with me on DH projects, mostly to discuss how to structure their project and to find out about resources beyond their own department. The events were well attended, especially by younger scholars and graduate students. After three years of building DHMS, from 2016–19, it is now in the capable hands of a younger colleague. While the DHMS initiative is far from normalized—given that collaborating units like the library or tech access remain in flux—I am much in favor of depersonalization as one faculty or staff should not dictate the course of

---

20   See 'DHMS Talk: Alan Liu, "Toward Critical Infrastructure Studies"', 1:32:20, posted online by University of Connecticut Humanities Institute, *Youtube* (August 18, 2017), https://www.youtube.com/watch?v=2ojrtVx7iCw&ab_channel=UniversityofConnecticutHumanitiesInstitute

21   See, e.g., https://dhmediastudies.uconn.edu/professional-links/ and https://dhmediastudies.uconn.edu/projects/

22   See https://dhmediastudies.uconn.edu/dhms-graduate-certificate/

an institute program that promotes collaboration in the first place. I also conducted a survey within the humanities and social sciences units that, unsurprisingly, confirmed what I had observed in the first year. With 50% of the respondents signing in as graduate students, most cited the lack of technical skills or time to embark on digital scholarship (71%) and a great need for workshops and seminars (69%) in addition to tech support. The response to 'what do you think is the future of digital scholarship in your field?' was positive, with some 'meh' or 'not sure' sprinkled in. One response summed up the general sentiment seeping through the survey results: 'bivalent bs: do digital humanities but still produce a book for promotion'.

For a complete institutionalization, directing an initiative such as DHMS should be a full-time position, tenured or tenure-track, and with an advisory board that reflects the resources and networks necessary to support a nascent community of digital scholars. Importantly, networking beyond one's own institution is key. DHMS's and therefore UConn's representation in a couple of regional DH networks is ongoing, namely the New England Humanities Consortium (NEHC) DH network I founded as a group affiliated with the Mellon-funded NEHC network originating from the UConn Humanities Institute; the Connecticut DH network I co-founded with a number of institutional representatives in the state; and I co-founded a new network within a discipline-specific organization, namely the DH Network at the German Studies Association.

From an advisor's perspective, directing DHMS has been quite successful as I can guide those students who work with me to utilize all available resources and begin to build their own networks. A graduate student who completed the DHMS certificate in 2020 published her digital scholarship in a peer review journal, was invited to present her work at a prestigious conference and ranked among top candidates for a DH position. Collaborating with other dissertation advisors should also help to build new networks, share knowledge and skill sets, and support graduate students in becoming digital scholars. However, it will take time, money and merit before digital scholarship at the dissertation level becomes fully institutionalized, at least judging from my vantage point at a large public, research one institution. We, as advisors, need this time, money or merit. As senior or tenured professors and as advisors, we are required to update curricula and integrate digital scholarship into the dissertation process: digital scholarship is part and parcel of humanities and art scholarship—who are we NOT to train our graduate students to be at least conversant in it, at the very least for career diversity and, ideally, for creating new epistemologies? Conversely, the university should create a central unit, either in the library or a humanities institute or DH lab, that becomes the go-to meeting place, exchange hub or brainstorm space to begin digital scholarship at any level and for a variety of purposes. As an advisor, I could send a student there should I not know how to advise her or him otherwise. And each PhD granting institution that has not established itself within DH cultures and aspires to a Duke or Michigan State or Northeastern or Brown or USC and many more, should give those professors credit, time, money or merit (preferably a combination thereof), for familiarizing themselves with digital scholarship such that they can train their graduate students for the twenty-first century. In Marshall McLuhan's *The Gutenberg Galaxy*, we read:

> The Gutenberg Galaxy is concerned to show why alphabetic man was disposed to desacralize his mode of being.[23]

It is time to desacralize traditional modes of academic being to allow for career diversity and experiments in knowledge production.

Bibliography

Antonijevic, Smiljana, *Amongst Digital Humanists. An Ethnographic Study of Digital Knowledge Production* (London: Palgrave/Macmillan, 2015), https://doi.org/10.1057/9781137484185

Cohen, Dan, 'Institutionalizing Digital Scholarship (or Anything Else New in a Large Organization)', *Dan Cohen* (November 29, 2017), https://dancohen.org/2017/11/29/institutionalizing-digital-scholarship-or-anything-else-new-in-a-large-organization/

Drucker, Johanna, 'Why Distant Reading Isn't', *PMLA*, 132.3 (2017), 628–35, https://doi.org/10.1632/pmla.2017.132.3.628

---

23   McLuhan, Marshall, *The Gutenberg Galaxy: The Making of Typographic Man* (Toronto: University of Toronto Press, 1962), p. 69

Drucker, Johanna, and Patrick B. O. Svensson, 'The Why and How of Middleware', *Digital Humanities Quarterly*, 10.2 (2016), http://www.digitalhumanities.org/dhq/vol/10/2/000248/000248.html

Fuller, Matthew, 'Nobody Knows What a Book Is Anymore', *Urbanomic* (2017), https://www.urbanomic.com/document/nobody-knows-book/

Kick, Verena, '"A Good Dissertation Is a Done Dissertation"—and Nothing Else Matters?', *HASTAC* (March 6, 2017), https://www.hastac.org/blogs/vkick/2017/03/06/good-dissertation-done-dissertation-and-nothing-else-matters

Kuhn, Virginia, 'Embrace and Ambivalence', *Academe*, 99.1 (2013), https://eric.ed.gov/?id=EJ1004358

Kuhn, Virginia, 'Multimodal', in *Digital Pedagogy in the Humanities: Concepts, Models, and Experiments*, ed. by Rebecca Frost Davis, Matthew Gold, Katherine D. Harris and Jentery Sayers (New York: Modern Language Association, n.d.), https://digitalpedagogy.mla.hcommons.org/keywords/multimodal/

Mattern, Shannon Christine, 'Evaluating Multimodal Work, Revisited', *Journal of Digital Humanities*, 1.4 (2012), http://journalofdigitalhumanities.org/1-4/evaluating-multimodal-work-revisited-by-shannon-mattern/

McLuhan, Marshall, *The Gutenberg Galaxy: The Making of Typographic Man* (Toronto: University of Toronto Press, 1962).

Presner, Todd, 'How to Evaluate Digital Scholarship', *Journal of Digital Humanities*, 1.4 (2012), http://journalofdigitalhumanities.org/1-4/how-to-evaluate-digital-scholarship-by-todd-presner/

Price, Kenneth M., and Ray Siemens, eds (2013-present), *Literary Studies in the Digital Age. An Evolving Anthology* (New York: Modern Language Association), https://dlsanthology.mla.hcommons.org/

Jeffrey T. Schnapp, *Knowledge Design* (Hannover: VolkswagenStiftung, 2014), http://jeffreyschnapp.com/wp-content/uploads/2011/06/HH_lectures_Schnapp_01.pdf

Spence, Paul, 'The Academic Book and Its Digital Dilemmas', *Convergence*, 24.5 (2018), 458–76, https://doi.org/10.1177/1354856518772029

Townsend, Robert B., 'Are Historians Still Ambivalent about Getting Published Online?', *History News Network* (2018), https://historynewsnetwork.org/article/168871

6. Findable, Impactful, Citable, Usable, Sustainable (FICUS)

A Heuristic for Digital Publishing

Nicky Agate, Cheryl E. Ball, Allison Belan,
Monica McCormick and Joshua Neds-Fox

Introduction

This chapter addresses some unanswered questions raised in this volume—primarily, how does one create a piece of digital scholarship that will be accessible and sustainable far into the future, if indeed that is a key component of the work (i.e., it is not event- or performance-based, or purposefully meant to be unarchivable). The authors of this chapter serve as digital scholarly experts—we are authors, editors, publishers, project managers, project directors and librarians for many digital journals, monographs and publishing programs; of individual, collaborative and cross-institutional digital humanities projects; and of digital publishing platforms being built to accommodate both large- and small-scale digital projects such as digital dissertations.

We came together in Spring 2018 at a two-day think tank hosted by Duke University Libraries and supported by The Andrew W. Mellon Foundation, with dozens of other librarians, publishers and scholarly communication stakeholders, to work on the question of sustainably publishing large digital projects. The outcome of that discussion turned into an extended project culminating in the heuristic presented at the

https://doi.org/10.11647/OBP.0239.06

end of this chapter. What leads up to that heuristic is how we created it and why it matters to your digital (dissertation) project.

Tending the Seeds of Sustainable Digital Projects

There is much research published in this book and elsewhere on the long (often unknown) history of digital scholarship, and the authors of this chapter have dedicated a good bit of their careers, in various work capacities mentioned earlier, to maintaining and creating sustainable workflows and platforms for archiving digital scholarly products— whether they are digital articles, monographs, journals and electronic theses and dissertations (ETDs); digital humanities projects that fall outside the scope of traditional peer-reviewed publications; or the platforms used to distribute and preserve these monographs, venues and projects. We know from our daily practice as digital librarians and digital publishers that the question of sustainability is not easily answered when it comes to working with scholars who desire to use the latest, greatest tools. There is often a tension between the use of innovative media and preservation of the scholarly projects it enables. It is disheartening when scholars spend hundreds of hours on a project, only to discover too late that the platform they have chosen doesn't afford them the chance to ensure the long-term viability of their work. This can happen for a myriad of reasons, including technological ease, existing knowledge base, accessibility, availability, economy and institutional constraints. The project wasn't built to be a performance piece, but it becomes one—a work slipping quickly into technological degradation and *un*planned obsolescence—because no one thought to consider sustainability as it was developed. We've seen entire scholarly journals disappear into the internet ether, including those managed and published by esteemed scholarly organizations.[1] But most publishers, librarians, editors and authors don't wish for that to happen.

That was the exigence for the two-day discussion at Duke, which raised questions about digital publishing workflows, from creation to preservation, for 'expansive' digital projects. These were defined as

---

1   Douglas Eyman and Cheryl E. Ball, 'History of a Broken Thing: The Multi-Journal Special Issue on Electronic Publication', in *Microhistories of Composition*, ed. by Bruce McComisky (Logan, UT: Utah State University Press, 2015), pp. 117–36.

digital humanities projects that are monograph-ish in scope. Many of the workshop participants were university press-affiliated publishers who have created or who manage publishing platforms that authors use to build digital humanities projects, including digital dissertations. Some of the most well-known of these open-source platforms (some of whose developers were in attendance) included Editoria, Fulcrum, Manifold and Scalar, but the group also had knowledge of authors who used other platforms such as Omeka and WordPress. The goal of the two-day workshop was to gather ideas on how a library should support authors who want to publish expansive digital projects, with the underlying issue being that many university presses—as the assumed go-to for many digital humanities (DH) authors—don't have the capability and/ or interest to offer long-term solutions for authors and their projects, whereas libraries are often better suited to help authors at most any stage of the DH project timeline and are *the* place where digital dissertations will eventually be deposited. (To be clear, the discussion at the workshop was *not* centered on graduate students and digital dissertations, but this chapter assumes that the digital dissertation is often the first type of project an author will undertake before embarking on a longer career filled with expansive digital projects and, indeed, they are often one and the same project as ETD grows into an academic's first expansive digital project post-graduation.)

While we authors represent a small fraction of attendees at the Duke Libraries workshop, it became clear from the discussion that the five of us[2] shared similar insights and expertise in publishing and preserving digital scholarly projects. At the end of the workshop, we were prompted by Paolo Magnifico (from Duke Libraries and project director for the annual Triangle Scholarly Communications Institute) to propose a working group for that year's TriangleSCI, where we would create a giant checklist/heuristic[3] for digital scholarly publishing that brought

---

2   The initial group included Melanie Schlosser from Educopia, but when it became evident that our discussions would extend beyond our time at Duke Libraries, Melanie (whose work availability was already structured so that she would miss some of our key meetings) suggested we bring Joshua Neds-Fox onto the team as an excellent library publishing representative.

3   We call the FICUS list both a checklist and a heuristic at different points, as it does the work of both: authors can use it to check off processes they have completed and can also use it to suggest ways of thinking about their projects that prompt actions towards findability, citability, etc. Therefore, we use the terms *checklist* and *heuristic* interchangeably throughout this chapter.

together existing best practices for publishing and preserving digital projects. There were and are many best practices, and more published regularly.[4] Our team was accepted for the week-long workshop in Durham, NC, to create what would become FICUS: a checklist for Findable, Impactful, Citable, Usable and Sustainable digital scholarship.

Fertilizing FICUS

A good DH project often starts with a catchy name, and the acronym FICUS came to us quickly. We appreciated that any checklist we made would be beholden to change—always in need of updating as new types of projects, technologies, genres and workflows were created around and in support of digital publishing. It made sense that our name reflect this precarity, and as sometime-gardeners, we recognized how precarious ficus plants are in the wild, easily dropping leaves and dying when environmental conditions shift. And yet they are beautiful, life-giving things. Our intention in naming the checklist after the ficus plant, then, is to indicate its usefulness while still understanding that the items within may change on a whim.

Our vision statement for the FICUS checklist highlights the necessity that 'digital projects are fully integrated into the scholarly publishing ecosystem and are recognized and rewarded as first-class scholarly contributions'. The 'first-class' designation came in response to then-Senior Program Officer at the Mellon Foundation Don Waters's use of the phrase to signal scholarly projects that are accorded the highest level of recognition and value in academia's tenure systems. That is, we wanted to mirror the language of one of the primary funding agencies to support digital humanities scholars and their work and to show that we firmly believe DH projects *are* first-class scholarly contributions within academia.

Our mission with FICUS is to 'reduce the risk for publishers by increasing the likelihood that digital projects will be findable, impactful, citable, usable and sustainable by building a scaffold of critical guiding questions'. The Duke workshop focused on library

---

4   See, e.g., Roxanne Shirazi and Stephen Zweibel, 'Documenting Digital Projects: Instituting Guidelines for Digital Dissertations and Theses in the Humanities', *College and Research Libraries*, 81.7 (2020), https://doi.org/10.5860/crl.81.7.1123

publishers, as evidenced by their 2019 outcomes publication,[5] which covered planning, allocating resources, discoverability, evaluating and preserving 'expansive' digital projects from a library's business-model perspective. At TriangleSCI, the FICUS team also decided to focus on educating publishers (including libraries) who wanted to help authors with digital projects. Our efforts later in this chapter turn this checklist towards authors—including those working on digital dissertations—and the information they need to plan and draft their projects, in consultation with their local librarians, potential publishers, and, of course, their advisors.

As we began to build FICUS, we drew from a number of existing resources that we and other participants at the TriangleSCI workshop knew about. It is likely there are even more resources that have become available since we first began work on FICUS. These resources are excellent sources of information on digital publishing in and of themselves, so we link to and explain them briefly here. It is basically the literature review section of this chapter. If you are ready for the checklist already, skip ahead to the next section.

An Ethical Framework for Library Publishing, Version 1.0: the 'Ethical Framework' (2018), authored by a working group of the Library Publishing Coalition that included Joshua Neds-Fox, a FICUS author, provides a heuristic for ethical considerations in digital publishing regarding accessibility; diversity, equity and inclusion; privacy; academic freedom; and related topics. The FICUS group focused primarily on the accessibility recommendations in this document to inform the usability and sustainability sections of our checklist, but approached the overall creation of the checklist in terms of its ethical role in helping publishers and authors to create projects that hit all possible marks for readership.

HuMetricsHSS Initiative: HuMetricsHSS began as a TriangleSCI project in 2016, where the project team created an humane values framework for 'evaluating all aspects of a scholarly life well-lived'.[6] These values include equity, openness, collegiality, quality and community. Nicky Agate, from the FICUS team, also serves on the HuMetricsHSS initiative and brought the concept of openness, in particular, to play

---

5   D. Hansen et al., *Expansive Digital Publishing* (2019), https://expansive.pubpub. org/

6   See https://humetricshss.org/our-work/values/

throughout our work on FICUS, and this work was particularly useful as we crafted the Impact section of the checklist.

'Access/ibility: Access and Usability for Digital Publishing': this 2016 publication on access and accessibility, openness, preservation and sustainability of digital scholarship came out of a weeklong workshop hosted by Cheryl Ball, one of the FICUS authors, and attended by twenty-six scholars, librarians and digital scholarship advocates. During the workshop, they created a set of best practices for accessible scholarly multimedia, built in part on the decades of experience publishing *Kairos*, the longest continuously running scholarly multimedia journal in the world. This list targets authors and publishers, and focuses on layout and design, interactivity, images, audio and video. The items here were primarily used for the Citable, Usable and Sustainable sections of the FICUS checklist.

DH Project Questions: this heuristic was created by FICUS author Ball to help authors translate some of the more challenging rhetorical and technical obstacles authors face when creating digital humanities projects into simple action-based questions they could answer. The list came from years of practice with *Kairos* authors and KairosCamp institutes where *Kairos* editors helped individual and collaborative author groups scope, pare, and propose better, more sustainable and rhetorically sophisticated digital publishing projects.[7] While this heuristic focused on authors and the FICUS team ended up focusing on publishers, some of the 'Big Questions' from this list, including 'Where will [your project] live?' and 'Who will sustain it?', guided how we created different categories of our FICUS checklist.

CRediT (Contributor Roles Taxonomy): CRediT provides a taxonomy of fourteen roles that represent the range of contributions often found in digital publishing projects, including Conceptualization, Data curation, Formal Analysis, Methodology, Project administration, Software, Visualization, Writing—original draft and Writing—review and editing, among others. The FICUS team used the concepts from CRediT to inform parts of the Impact section, particularly as it relates to tenure and

---

7   A more contextual version of these questions is forthcoming in Eyman and Ball's chapter 'Everything is Rhetoric: Design, Editing, and Multimodal Scholarship', in *Editors In Writing: Behind the Curtain of Scholarly Publishing in Writing Studies*, ed. by Greg Giberson (Logan, UT: Utah State University Press).

promotion/evaluation issues (i.e., *who* gets credit for working on digital publishing projects and how are those folx' work rewarded?).

NDSA Levels of Digital Preservation: the National Digital Stewardship Alliance (NDSA) has provided a matrix for digital preservation of all *kinds* of projects since 2013, and their updated 2018 version was in-progress at the time we were working on FICUS, but still provided a roadmap for parts of our Sustainability section, in particular. Their matrix provides different levels of focus on preservability for libraries and archives to follow that focus on knowing, protecting, monitoring and sustaining one's digital content.

FAIR data principles: these principles are targeted towards making data-intensive science and data sets more Findable, Accessible, Interoperable and Reusable (FAIR). They were published in 2016, but they didn't come to our attention until the 2018 TriangleSCI workshop, thanks to a group of our European colleagues (where the original principles were created). The FICUS group noted the cross-overs between both sets' Findability and Accessibility principles, and that much of what the FAIR principles outline in terms of data can easily be applied to digital publishing projects writ large.

Socio-Technical Sustainability Roadmap (STSR): this project, hosted by the University of Pittsburgh's Visual Media Workshop, was published while we were at TriangleSCI and covers a broad range of sustainability questions for digital projects. The FICUS team felt that the STSR was more comprehensive in covering some of the sustainability issues than we could cover in a week of brainstorming, so our Sustainability section remained in beta until writing this chapter. We still refer publishers and authors to that document, particularly as it highlights questions in regards to a whole project (the questions of which are similar to the DH Project Questions discussed above), staffing and technologies, and creating a digital sustainability plan for projects.

'Developing a Business Plan for Library Publishing', by Kate McCready and Emma Molls, was published in 2018, around the same time as our TriangleSCI meeting. Although we didn't use it to inform our FICUS checklist, the concept of providing guiding questions to establish an effective and sustainable library or other publishing program will impact the sustainability of digital projects that any publisher undertakes, and that authors should be aware of. Ultimately (and, in our minds, unfortunately), the business models of a publisher will indicate and often limit the types of projects publishers are willing to move forward with.

Tending FICUS

The questions in the Findable, Impactful, Citable, Usable, and Sustainable sections provide a framework for authors. Initially geared towards publishers, we have transformed the FICUS checklist to accommodate how authors of digital projects might use these questions. This checklist doesn't follow a linear order for composing a digital project, however. Authors might want to start with the Usable section, as it provides an entry point for applying your project's purpose to an audience in a usable fashion. Next, authors may want to review questions in the Sustainability section, because it outlines how to find the best platform for your content and how to prepare your text for longevity from its earliest beginnings. From there, we recommend working backwards through Citable, Impactful and Findable, as many of the questions in Findable will require interaction with a librarian or publisher. While we don't have space here to annotate each of the questions, and we recognize that some of the questions may more firmly rest in the domain of publishers and/or librarians, they are still good questions for authors to discuss with their librarians/publishers to learn more about the publisher's approach to the longevity of their digital scholarly projects. The sections that have publisher-relevant questions are marked as such by the header to 'Ask your Librarian and/or Publisher'.

In the case of ETDs, which are often published via an institutional repository, the key is to find the person(s) in your library or digital humanities center or research office who may be called a scholarly communications librarian, digital scholarship librarian, copyright officer, institutional repository manager or other titles (which may not include the word 'librarian'). Look for the person who has some familiarity with digital publishing and can help you navigate these questions. Indeed, as you get deeper into the FICUS checklist, we hope it will become more obvious that partnering with a digital librarian means more than just relying on them to answer some basic questions for you, but that these folx can be embedded in your project team from the beginning and will often contribute a great deal of intellectual labor to your project. The CRediT taxonomy discussed earlier offers suggestions for how to credit them. TaDiRAH, the Taxonomy of Digital Research Activities in the Humanities, can also help in this regard. (But, y'all, for the love of all things easily citable, do we NEED these random capitalizations?!) We also recognize that not everyone will have a digital librarian at their university to ask these questions, and that shouldn't stop you from proceeding! Start by asking your advisors or dissertation committee members, and if they don't have any experience with digital dissertations and these types of questions, you might reach out to other scholars on your campus or other digital dissertators you know. In any case, coming to these allies having thought through possible answers is a great strategy for getting them up to speed on your project, so they will understand the scope of your needs and desires. Librarians and publishers in particular will do their best to help you fulfill both of those, though compromise is often required in digital projects due to usability, accessibility, sustainability, economy, and availability issues.

FICUS (the Checklist)

Findable

This section helps authors answer the question: how findable is your project, both by humans and by machines?

Ask Yourself and/or the Project Team

- Does the project fit (disciplinary, subject, methodological) into an existing publishing venue, index, list, or aggregation?

- Who is responsible for promoting and publicizing the project, and what methods will be used to do so?

- Where does your target audience discover new scholarship?

- If your project is about a certain ethnic, racial, geographic, socioeconomic group, how will you ensure that those audiences know about it?

- Can users in other languages/countries/environments discover the project?

- What partnerships can you form or use to create awareness of the project?

- What venues review projects like this?

Do you have an ORCID (http://orcid.org) and other persistent social media handles that will link you to the project once it is published?

Are commercial search engine optimization techniques employed for the project?

Does the project need to incorporate linked open data to enhance discovery and use, and how will you or your publisher provide for that?

Ask Your Librarian and/or Publisher

- What metadata needs to be created/maintained in order to register this project with the appropriate discovery systems?

- Does your metadata schema enable web-scale discovery? Specialty system discovery (e.g., library OPAC, DPLA, etc.)?

- What other persistent identifiers (work, object, media, personal) are relevant and/or useful to the project? (DOI, ORCID, ISSN, ... more?) Do those identifiers support the kinds of objects, media, work, persons involved in your project?

- Which of the following will the identifiers and registries provide (note that not all are required, but you should consider which your project requires):

  o Unique ID

  o Persistent link

  o Associated metadata repository/registry

  o API access to the repository/registry to services (such as reference linking, reference lookup, interaction with other services—funder repositories, for example)

  o Identity disambiguation

  o Credit

- Is there a plan for maintaining the project's metadata in the identifier registries?

Impactful

This section answers the questions: will your project have impact and how will it be assessed?

Ask Yourself and/or Your Project Team

- Does this project fit with the broader goals of your academic research or teaching trajectory (e.g., scholarly/disciplinary focus, technology use, institutional mission/vision/goals)

- Will the project or its participants need or benefit from a scholarly assessment and validation process? (for validation, for tenure and promotion)

  - What form of scholarly assessment and validation is most appropriate for the project? (pre-publication review, post-publication review, open review, anonymous review)

  - Who is responsible for conducting the scholarly assessment and validation process? (e.g., the authors, the project team, the publisher)

  - Will the project document its scholarly assessment and validation method?

  - What do stakeholders need to know about the scholarly assessment and validation process for this project?

  - At what stages of the project will it be subject to scholarly assessment and validation?

- Who is recognized as a contributor to the project and how is that recognition expressed (human readable, machine readable)?

- How will you design the project to ensure that all project partners' valued metrics are captured?

- How will you measure the success and impact of the project?

  - How will use be measured (course adoption, inclusion in LibGuides, downloads, web traffic, time on page)?

    o How will engagement be measured? (i.e., citations, blog posts, annotation, reviews, discussion in news media, assignments, community interest)

    o How will impact be measured? (i.e., international reach, awards, inclusion in public policy documents, references in grant proposals, citations, inclusion in syllabi)

- Is the project designed such that the desired measurable outputs can be tracked?

- Is the project designed such that its various uses can be tracked and followed?

Ask Your Librarian and/or Publisher

- Does this project fit your technological profile, either existing or aspirational?

- How will this project enable you as a publisher to broaden or deepen the scope of what you can offer?

- Does this project illustrate or demonstrate your values?

- Will you conduct scholarly assessment and valuation of the project, and if so, how will that be documented?

- How might you help us measure the project's use and impact?

Citable

This section answers the question: how, and in what forms, will your work be cited by other scholars? All of these questions might best be answered in consultation with a digital librarian.

- Given the nature of the project and its content, what is the unit of scholarly value that users will want to cite? (e.g., the entire project, pages/sections/units within the project, individual media assets within the project, etc.)

- Does the project have the markers of permanence (including persistent identifiers) that make scholars secure in citing it?

    o If there is more than one citable unit, does each have a persistent identifier?

- Is integration with automatic citation generators desired or possible?

- How does project type/media/genre affect what's citable, the citation format that may be used, and the metadata required?

- If the project in its public form changes over time, what is the plan for maintaining citability?

- If the content is later edited or modified, how will you ensure that 'version control' is reflected in the citation?

Usable

Usability is a more encompassing issue than the previous sets of questions, so we've broken it down into sections that each address different aspects of making your project usable. You will likely want to consult with a digital librarian on most of these sub-sections.

Audience

Answers the questions: is your project internally coherent in regard to its intended audience? How does your intended audience drive your choices for technology, language, design, etc.? This section is not meant to address the entire rhetorical scope of how audience affects your digital project, but to address how audience and usability intersect in terms of creating sustainable projects.

- Who is/are the audience(s) for your project? Is the project's audience well-defined?

- How does the platform choice impact the potential audience's use of the project? (See also *Sustainable*)

- Is there interaction with the project? Will that interaction be public, in the form of community translations, annotations, comments, or contributions? Will they be instantly visible, or after moderation? If so, how will that mediation or moderation take place and who will do it? (See also *Sustainable*)

- Are you attempting to crowdsource any part of the project content? How? (See also *Sustainable*)

- Will the intended audience have the necessary technical expertise and affordances (e.g, infrastructural access)?

- Does the project discovery plan serve the intended audiences? (see also *Findable*)

- Does your project's development plan allow for the discovery and accommodation of unexpected audiences? (see also *Findable*)

- How will you determine that your project is reaching its intended audience, or recognize other audiences it is reaching?

Accessibility

Answers the question of who has access to your project, focusing on people with physical, geographic, and economic barriers to access.

- What is the accessibility testing plan? (timing, frequency, stakeholders, target compliance levels)

- Will the project be accessible on different devices?

- Who is included in usability testing and is that group inclusive of people of differing abilities and backgrounds?

- What statutory or institutional guidelines or requirements is the project subject to?

- How will the project's device and browser support impact the expected and unexpected audiences' access?

- How will the project's content, context, and structure allow or limit access outside of its geo-political and cultural context? (bandwidth, reliability, expense, language, software, graceful degradation, social accessibility/censorship)

- Does the project allow for effective, authentic access to critical stakeholder communities? (e.g., those whose work or communities are featured in the project)

Usability

This section answers the question of how usable your project will be to potential audiences. You may develop a usability testing plan in concert with your library or other publisher.

- What is your usability testing plan?

- Is access limited by IP/username & password/non-accessible platform/language? (See also *Sustainable*)
- Have you developed testers that mirror the project's audience?
- Will the project's content be understandable in either human or machine-readable ways when encountered outside of the designed application?
- Have the design and layout elements of the project been assessed for loss of meaning if they are removed, absent, or do not gracefully degrade? (see also *Accessibility*). For example:
  - Will the text still function if a user views the project with their own style sheet?
  - Is navigation available through multiple modal points (e.g., mouse, trackpad, keyboard, eye-tracker, etc.)?
  - Do user interactivity features include feedback mechanisms, such as confirmation of response, indication of progress toward completion, time left to complete or timeout, etc.?
  - Do all media assets (image, audio, video, etc.) have attached descriptors and proper structured text (e.g., transcripts, captions, descriptions, alt attributes, etc.)?
  - Do users have control over how media assets are to be interacted with? (e.g., turning off auto-play on videos, etc.)? Do animated/moving assets avoid rapid refresh rates, blinking, pulsing or quick movement of dots and narrow stripes?
  - If color were removed, would the project's use be inhibited?

Intellectual Property and Use Rights

This section answers the questions: How does copyright and licensing work in and for your project and team members? Some answers may be dependent on your publishers' requirements as outlined in their author agreements, so you may need to address these questions in consultation with them.

- Will individuals keep copyright of their individual contributions? Will teams collectively share copyright to the outputs?

- What license will be applied to the project (all rights reserved; open license such as CC BY, CC0, GNU, WTFPL, EUPL)?

- Will different licenses be applied to different parts of the project (metadata, software, data)?

- Are there institutional policies that may guide or constrain your licensing options?

- Will the license choice impact the project's eligibility for inclusion in relevant aggregations, indexes or other third-party discovery systems?

- Does the licensing structure support the intended uses and appropriately restrict other uses?

- Is the license both machine and human readable?

- Does the project include works/assets that are under copyright or require a license to use?

- Are there licensing limitations (use, cost, format quality) that would negatively affect the usability, accessibility, sustainability or impact of the project, either now or in the future?

- What materials fall under fair use? Public domain?

- How will copyright and credit be acknowledged or attributed in the project?

- What parts of your project are intended for reuse (content, data, platform, etc.)?

- What modes of technical re-use are intended? (replicable, consumable, portable)?

- How does the design and structure of the project allow for intended re-uses (e.g., package, zip file, Docker, Vagrant, GitHub, API, etc.)?

    o If the content layer (separate from the structure) is meant to be re-usable, how does the project accommodate that (e.g., APIs, OAI-PMH, data portability through structured content using json, XML, etc.)?

- o Are underlying systems essential to the project's re-use? (programming environments/languages, dependencies, software, hardware, operating systems, etc.)

- Does the project make use of descriptive standards that promote its re-use? (e.g., metadata schema, rational URLs, Persistent ID systems, etc.)?

- Will the programming or content language be a barrier to re-use for your intended audiences?

- How will the project prevent unauthorized reuse of restricted materials?

- How does the project's copyright status or license impact its reuse?

Sustainable

The Sustainability section relies heavily on the work of the Socio-Technical Sustainability Roadmap and the NDSA Levels of Digital Preservation. This section answers the questions: what content do you have and what platform will you need? How will you work with these materials to make the project sustainable? What is the end-life of your project?

Platforms

This section provides questions to help you determine a platform given your content, audience and tool availability, and how that platform will be maintained. A list of digital tools that have been used for DH projects is given here, although the maintenance of the list is in question, as new platforms, technologies and the like change rapidly: https://digitalhumanities.berkeley.edu/resources/ digital-research-tools-dirt-directory

- What is the project for? (Is it a house? A power tool? A community? An attic?)

- Is it meant for people (to use, view, act on, work with) or not (does it operate on things by itself)?

- Is the project static or dynamic?

- How will the project interact with people or other systems? Do either of these need to add to/alter the project? Does it need to maintain states?

- Do you need user management/proxy identities?

- Is there some skill/knowledge/ability required before one could engage this project? (programming language, disciplinary knowledge, technical affordance) Does the platform need to mediate that Knowledge, Skill, Ability (KSA)?

- What content types live in the platform? (data, images, text, software, video, audio, complex digital objects, metadata, a stream of content from somewhere else, something else)

- Does the system need to manage persistent identifiers for content? (See also *Findable*)

- Is it meant to be open source or proprietary?

- What computing power is needed for the project? Is it resource intensive? (grid power, CPUs, memory use)

- How much digital space will this take? Will the project grow/ shrink?

- Does the project need a human/s to manage data, software/hardware, development, workflows, users?

- How much institutional/ideological support and enthusiasm does the project have? Will the project die without you?

Preservation

This section answers the question of how long your project needs to last and how it will be preserved.

- How long does the project need to last to serve its purpose?

- Does it need to remain usable/reusable?

  - Is an analog, abstract, report, record or snapshot of the project sufficient for the long term?

- Are the systems/formats used in the project integral to the nature of the project, or could it be migrated to a new system/format if current systems become obsolete?
    - Are the modalities migratable? Is the user interface integral to the project, or could it be reconceived?
    - Will the formats of the project degrade physically? Is there a storage/migration solution for these formats if so?
    - What systems also need to be preserved in order to ensure long-term viability of the project?
    - Do you have access to the infrastructures necessary to preserve those systems?
- Is there sustainable funding earmarked specifically for preservation?
    - How much?
    - Will/can the project generate income? If so, is this income enough to solely sustain the project for its entire lifespan?
- Does the project have or need policies to describe the preservation intent (to protect it against commercial capture, commercialization, and/or disintegration)?
- How much digital space (GB/TB/PB?) does the project occupy?
    - Is there a second, geographically separated digital place where that much space can be apportioned for a copy of the project?
    - Has this space been budgeted for in a sustainability/funding plan?
- Can the storage/preservation versions of the project or its content be reliably reconstructed? How do you determine whether the data has degraded over time (checksums, etc.)
- How stable is the institution the project is connected to? Is it likely to last? Are there political considerations to the longevity of the project?

- Do natural, geophysical or geopolitical realities threaten the long-term viability of the project?

Retirement

Answers the questions of how to plan for your project's digital afterlife.

- Once it goes live, is it finished/final/complete/closed to ongoing work *or*
- Is it intended to be developed/augmented/expanded continuously?
    - o How will you ensure that work continues? On what cycle?
    - o By what metric will you measure 'fruitful' expansion?
    - o Will your publisher allow for ongoing work?
    - o How will you know when the project has realized all the value it can?
    - o Will there be periodic review of project value/viability? How often? By whom?
- How will you communicate the status of the project to users?
- What metrics will indicate that the project has reached the end of its lifespan?
- Is there a community that should be consulted with or communicated with about lifecycle events?
- When the project is at the end of its life, what constitutes adequate digital hospice? How do you help this project degrade gracefully into that good night?

FICUS in the Wild

It won't be surprising if authors read the FICUS heuristic with bewilderment at the depth of thinking, pre-planning and execution of minutiae that seems required of digital projects. Librarians get that, which is why it is literally our job (at some institutions) to help scholars think through these types of projects. It also wouldn't be surprising if authors only followed a small portion of these recommendations. It has been true for the decades since digital dissertations became objects of the scholarly record that authors have *not* attended to many of the items on this list, because these items felt outside your purview, beyond your knowledge or, in some cases, not even a *thing* yet (e.g., early ETDs that were published during a time when DOIs and rich metadata didn't yet exist). We get it, and we sympathize. Your intellectual contributions towards the growing content knowledge in your academic disciplines are still the primary consideration in your dissertation projects and the primary expectation of your committees; the technological considerations, out of which much of this checklist is built, have most often been used in service of the content, which means they are considered after-the-fact and with whatever technology was available at hand. Our goal with the FICUS list is not to provide a mandate of to-dos for every digital dissertator but a set of considerations that will make your intellectual and technological labor last far into the future, for many more researchers to engage with.

We have published an archived version of the heuristic for you to download and use as an actual checklist at https://digitalcommons. wayne.edu/libsp/152/. We will end here, then, with the possibilities that present themselves to you as you proceed in your research, and the hope that you might let us know how it is going if you use the FICUS list.

Bibliography

Agate, Nicky, et al., 'Findable, Impactful, Citable, Usable, Sustainable (FICUS): A Heuristic for Authors of Digital Publishing Projects', *Digital Commons*, https://doi.org/10.22237/waynestaterepo/libsp/1606953600

Ball, Cheryl E., 'DH Project Questions', https://docs.google.com/document/d/194VWvDG9PZbDA6GmIV-QsT3wZ9-UE0CQ5CLtXYGoWls/edit?usp=sharing

Boczar, Jason, et al., 'An Ethical Framework for Library Publishing', *LPC Publications* (2018), Paper 1, http://dx.doi.org/10.5703/1288284316777

CASRAI, 'CRediT—Contributor Roles Taxonomy', https://casrai.org/credit/

Digital Humanities at Berkeley, 'Digital Research Tools (DIRT) Directory' (2015), https://digitalhumanities.berkeley.edu/resources/digital-research-tools-dirt-directory

Eyman, Douglas, and Cheryl E. Ball, 'Everything is Rhetoric: Design, Editing, and Multimodal Scholarship', in *Editors in Writing: Behind the Curtain of Scholarly Publishing in Writing Studies*, ed. by Greg Giberson (Logan, UT: Utah State University Press, forthcoming).

Eyman, Douglas, and Cheryl E. Ball, 'History of a Broken Thing: The Multi-Journal Special Issue on Electronic Publication', in *Microhistories of Composition*, ed. by Bruce McComisky (Logan, UT: Utah State University Press, 2015), pp. 117–36.

Eyman, Douglas, et al., 'Access/ibility: Access and Usability for Digital Publishing', *Kairos*, 20 (2016), http://kairos.technorhetoric.net/20.2/topoi/eyman-et-al/index.html

FORCE11, 'The Fair Data Principles', https://www.force11.org/group/fairgroup/fairprinciples

Hansen, D., et al., *Expansive Digital Publishing* (2019), https://expansive.pubpub.org/

HuMetricsHSS, https://humetricshss.org/

McCready, Kate, and Emma Molls, 'Developing a Business Plan for a Library Publishing Program', *Publications*, 6.4 (2018), 42, https://doi.org/10.3390/publications6040042

NSDA, 'Levels of Digital Preservation, Version 2.0' (2019), https://ndsa.org//publications/levels-of-digital-preservation/

Shirazi, Roxanne, and Stephen Zweibel, 'Documenting Digital Projects: Instituting Guidelines for Digital Dissertations and Theses in the Humanities', *College and Research Libraries*, 81.7 (2020), https://doi.org/10.5860/crl.81.7.1123

TaDiRAH—Taxonomy of Digital Research Activities in the Humanities, http://tadirah.dariah.eu/vocab/index.php

Visual Media Workshop at the University of Pittsburgh, *The Socio-Technical Sustainability Roadmap*, https://sites.haa.pitt.edu/sustainabilityroadmap/

SECTION II

SHAPING THE DIGITAL
DISSERTATION IN ACTION

7. Navigating Institutions and Fully Embracing the Interdisciplinary Humanities

American Studies and the Digital Dissertation

Katherine Walden and Thomas Oates

In recent decades, new allowances for multimodal, digital forms of American Studies scholarship have been palpable. For example, members of the American Studies Association created a Digital Humanities Caucus in 2009, while its flagship journal *American Quarterly* recently expanded its reviews section to include assessments of digital projects, and announced plans for a special issue titled 'Towards a Critically Engaged Digital Practice: American Studies and the Digital Humanities', which was published in October 2018.[1] But while American Studies as a field has recently advocated for and accepted alternative forms of scholarship, many questions and uncertainties linger for PhD-granting programs faced with the prospect of credentialing new forms of scholarship as sufficient to meet the completion requirements for the subject area's terminal degree. For doctoral programs whose mission and vision maintains an enduring commitment to training future generations of the professoriate, moving into the unknown territory (and attendant uncertain career trajectories and prospects) of alternative forms of scholarship raises the following fundamental questions:

---

1   'Special Issue: Toward a Critically Engaged Digital Practice: American Studies and the Digital Humanities', ed. by Lauren Tilton et al., *American Quarterly*, 70.3 (2018).

    https://doi.org/10.11647/OBP.0239.07

- What is a dissertation?
- What are the core intellectual tasks and academic skills doctoral students must master to successfully complete a dissertation?
- Where does the dissertation fit in relation to evaluating a doctoral candidate's successful completion of the PhD degree program?
- How will investing in and advocating for digital forms of scholarship impact student prospects and competitiveness on the job market?

As evidenced by our experience with a born-digital American Studies dissertation project at the University of Iowa, graduate students and faculty interested in proposing and advocating for an alternative dissertation project can encounter significant obstacles in this process, even at campuses where there is institutional or administrative support for emerging forms of scholarship and new approaches to graduate education. This chapter addresses the steps and resources academic departments can take to effectively support and equip graduate students for completing large-scale digital projects. It highlights the challenges graduate students and faculty face in advocating for alternate dissertation forms, and addresses the institutional and intellectual challenges digital dissertations present, in order to shed light on the logistics of preparing for, undertaking, and completing a digital dissertation. Additionally, we explore how graduate students and faculty advocates invested in large-scale digital projects can utilize and leverage institutional resources, professional organizations, and other communities and networks to expand the possibilities for humanities dissertation forms.

The idea for 'Remapping and Visualizing Baseball Labor: A Digital Humanities Project' began in an Archives and Media course that, at the time, was a required course for the University of Iowa's Certificate in Public Digital Humanities. This course included a semester-long data management and visualization project. Through gathering, organizing, analyzing and visualizing a small sample of baseball-related data, and the research questions and areas of inquiry facilitated by that experience, the idea of a larger project based on a more complete data set emerged.

Multimodal Digital Content as Argument

Contemporary digital humanities (DH) scholarship includes vigorous calls for humanists to create meaningful contributions to scholarship. For instance, the authors of the *Digital Humanities Manifesto 2.0* envision the use of digital tools or resources to address core humanities methodological commitments like 'attention to complexity, medium specificity, historical context, analytical depth, critique and interpretation'.[2]

The scholars and projects featured in the 2018 *American Quarterly* special issue, among others, have marked American Studies as a versatile, interdisciplinary home from which can emerge DH projects that have the capacity to inform and shape ongoing scholarly conversations. American Studies adjacent projects such as *Digital Harlem*[3] and *Early African-American Film Database*[4] render textual data—whether text from a newspaper primary source or descriptive information for a film—in tabular form, visualized or graphically rendered in a way that illuminates compelling narratives and unexpected intersections. More than a mere illustration of written arguments, *Digital Harlem* utilizes newspaper accounts from the African-American press to map the spread of black cultural institutions in a particular geographic space in a way that traditional forms could not. *Early African-American Film Database* uses filmmaker and production information for a corpus of race films in a dynamic online site that includes humanities-oriented discussions of the dissertation's method and significance, as well as tutorials on a range of ways to interact with and visualize the archive. Both projects offer compelling examples of the utility of digital humanities scholarship and serve as models for interactivity and visualization within the field of American Studies.

While much DH work has utilized digital tools and employed digital methods to explore primarily textual sources, data visualization practices in journalism offer a rich body of examples that illustrate the potential value and utility of data mapping and visualization approaches to topics,

---

2   Humanities Blast, *Digital Humanities Manifesto 2.0*. (2009), p. 2, http://www.humanitiesblast.com/manifesto/Manifesto_V2.pdf

3   See http://digitalharlem.org/

4   See https://web.archive.org/web/20201109105514/http://dhbasecamp.humanities.ucla.edu/afamfilm

issues and questions of interest to humanities scholars..[5] In addition to advocating for digital alternatives to the traditional dissertation, this project seeks to illustrate how digital dissertations can move beyond textual studies and engage digital approaches and resources to expand the types of argumentation and knowledge communication central to the humanities dissertation.

The path, however, was more complicated than either of us anticipated.

Power of Precedent

The most immediate obstacles to the proposed project were logistical. The graduate college did not have a formal mechanism to support the deposit of a project that could not be manifested as a PDF. Fortunately, however, a Digital Studio for Scholarship and Publishing had been established, headed by Dr. Deborah Whaley, a faculty member in American Studies at Iowa and a member of this dissertation committee with expertise in DH projects. With Dr. Whaley's assistance, arrangements were made to house the final version of the dissertation. Even after resolving this fundamental issue, however, another logistical obstacle remained. The American Studies Department's graduate handbook described the dissertation stage of the program as follows: 'A PhD dissertation or thesis in American Studies is a substantive book-length manuscript that involves interdisciplinary research, analysis, and represents an original contribution to knowledge'.[6] Thus, we began a conversation among American Studies faculty about the possibility of changing the language

---

5   Martyn Jessop, 'Digital Visualization as a Scholarly Activity', *Literary and Linguistic Computing*, 23 (2008), 281–93, https://doi.org/10.1093/llc/fqn016; A. V. Pandey et al., 'The Persuasive Power of Data Visualization', *Transactions on Visualization and Computer Graphic, IEEE*, 20 (2014), 2211–20, https://doi.org/10.1109/tvcg.2014.2346419; John Theibault, 'Visualizations and Historical arguments', in *Writing History in the Digital Age*, ed. by J. Dougherty and K. Nawrotz (Ann Arbor: University of Michigan Press, 2013), pp. 173–85, https://doi.org/10.2307/j.ctv65sx57.19; Erik Malcolm Champion, 'DH is Text Heavy, Visualization Light, and Simulation Poor', *Digital Scholarship in the Humanities*, 32 (2017), i25–32 (at 25), https://doi.org/10.1093/llc/fqw053; Elijah Meeks, 'Is Digital Humanities too Text-Heavy?', *Digital Humanities Specialist, Stanford University Libraries* (July 26, 2013), https://dhs.stanford.edu/spatial-humanities/is-digital-humanities-too-text-heavy

6   American Studies at the University of Iowa, 'Guidelines for American Studies Graduate Students' (2017), p. 3, https://clas.uiowa.edu/american-studies/sites/clas.uiowa.edu.american-studies/files/handbook_revised_Fall17.pdf

in the handbook to facilitate digital projects. This petition for a more flexible definition of what could constitute a dissertation in the program led to a much larger conversation about the place of digital methods and digital scholarship in the graduate program, the appropriateness of alternate forms of scholarship for graduate students in the program, and larger ontological and epistemological questions about what constitutes a dissertation, writ large as well as within American Studies.

As we learned, graduate students, faculty advisors, department faculty and graduate college administrators face a number of challenges when trying to establish procedures or conventions for non-standard dissertation projects. The language that many academic organizations have produced, created or adopted around digital scholarship standards in relation to faculty promotion and tenure suggests there is some hope for articulating, with some degree of concreteness and clarity, the expectations for digital work. But, as Virginia Kuhn points out, the existence of those standards and guidelines often is of little material benefit to junior faculty in departments that are slow to fully adopt or implement those digital standards. Across disciplines, but particularly within the humanities, the dissertation is framed as a single-author scholarly project that represents a student's intellectual contributions, analysis and arguments. While the conversations about intellectual/academic labor and authorship clearly demonstrate that all scholarship is inherently collaborative, there is little precedent for a dissertation project that is conceived of and proposed as a collaborative project.[7] At least within humanities disciplines, the philosophies and assumptions about academic labor undergirding the dissertation make it challenging to see a clear path toward adopting or adapting guidelines for collaboration that can be applied to faculty scholarly work. Similarly, the logistical and administrative challenges of having external readers or members on a dissertation committee make it difficult to advocate for external peer review as a feasible model for evaluating digital dissertation projects.

As our discussions about the handbook language progressed, the value of precedents became clear. Specifically, studying peer institutions who have supported alternative dissertations can help those proposing

---

7   Whearty, Bridget, 'Invisible in "The Archive": Librarians, Archivists, and The Caswell Test', *Medieval(ist) Librarians and Archivists: A Roundtable*, 53rd International Congress of Medieval Studies, Kalamazoo, MI, May 10–13, 2018.

or advocating for new dissertation forms, since they have a fuller understanding of the administrative and institutional complexities and challenges these projects can raise, while they also point the way toward strategies for navigating such obstacles. The digital scholarship undertaken by faculty on our own campus as part of their research agenda also helped position a digital dissertation project as within the scope of legitimate scholarly activity happening on our campus.

The conversations happening in forums like HASTAC 2015's Remix the Diss panel, Amanda Visconti's personal blog[8] and other spaces are a useful starting place for gathering resources and information to advocate for, propose revised handbook language for, and evaluate alternative types of dissertation projects. Similarly, the handbook language used in fields or programs that accept alternative theses and dissertations can also be a useful starting place for developing a handbook language revision proposal, while it can inform discussions between a student and advisor when establishing the tangible deliverable components of a digital project. For example, the language that George Mason University's History and Art History Department adopted in 2015 for digital projects was the inspiration for the proposed handbook language that was eventually adopted by the University of Iowa's American Studies Department. As more professional organizations and academic departments come to terms with the reality of an increasing number of alternative and digital dissertation projects, we hope institutions will follow George Mason's precedent of making those materials available online.[9]

These conversations raised interesting intellectual and pedagogical questions, but they took months to play out. On a practical level, graduate students and advisors need to consider issues of degree timeline and time to degree when proposing digital projects. As anyone with experience undertaking digital projects or learning new digital skills can attest, taking a digital or nontraditional approach to a scholarly project is not a shortcut to a lighter workload, a less grueling dissertation, or a faster completion timeline. If the experience of this

---

8   See http://literaturegeek.com/tag/dissertation

9   Department of History and Art at George Mason University, 'Digital Dissertation Guidelines', https://historyarthistory.gmu.edu/graduate/phd-history/digital-dissertation-guidelines

project is an instructive model, the process of navigating department and administrative conventions and procedures, while also negotiating within a committee about what exactly the digital project will look like, makes the dissertating process more laborious and time-intensive.

Collaboration, Project Management and Single-Authorship

We believe that a fully transparent collaborative dissertation model is likely the major transformation or discussion that will follow the debates around alternative dissertation projects. However, in the current framework for humanities graduate training that usually requires single-authored dissertations, one of the key challenges an alternate dissertation project presents for doctoral students, graduate advisors, committees and departments is how to best support or facilitate the collaboration needed to acquire sufficient skills to undertake and execute an alternate dissertation project.

For some alternative dissertations, students come to the project with a pre-existing set of technical, digital, or creative skills, as in the case of Nick Sousanis' *Unflattening* (2015) graphic novel or A. D. Carson's *Owning My Masters: The Rhetorics Of Rhymes & Revolutions* (2016) album. For students who come to the dissertation stage with the skills necessary to execute an alternative project, the process of proposing and gaining approval will likely involve demonstrating and leveraging those existing skills and illustrating how the dissertation forms made available via those skills constitute a valid or substantive scholarly contribution.

However, as graduate schools and some graduate programs become increasingly invested in and committed to increasing graduate students' digital competency and capacity to communicate or disseminate their scholarship in multiple forms, the skills necessary to accomplish those goals can strain the limits of existing graduate curricula. More practically, those hybrid curriculum initiatives are more likely to succeed in equipping graduate students with those skills when they include cluster hires for tenure-track digital scholars. Otherwise the impetus to train graduate students with digital and multimodal skills comes up against the limitation of faculty teaching graduate courses who do not engage in that type of work in their own research practice, and who are often operating within a tenure and promotion structure that places greater weight on traditional forms of scholarship.

While widespread acceptance for collaborative dissertations has yet to be fully realized, graduate students, faculty, and institutions can identify and make available institutional resources that are able to provide students with opportunities to gain the additional skills necessary to execute or even imagine an alternative dissertation project. The digital dissertation highlighted in this article began in a Library and Information Science course that was a required component of a Public Digital Humanities Certificate program. From an introduction in digital humanities and digital pedagogy to more specialized technical training, the Certificate coursework at the University of Iowa is one initiative that formalizes institutional partnerships and affiliations to identify the faculty, departments, and courses that can support students interested in digital or alternative forms of scholarship, whether they be more interested in digital pedagogy or a full-scale alternative dissertation project.

In addition to collaboration as a means of skill acquisition, forming connections and relationships with other campus units that will be able to provide technical, infrastructure and preservation support is also a necessary step for the long-term stability of digital dissertation projects. While cloud or site-hosted programs like WordPress, Omeka and Scalar have lowered the barriers of entry for scholars interested in pursuing alternative forms of scholarship, the technical infrastructure needed to carry out a digital dissertation likely moves beyond the capacity of many site-hosted programs, and depending on the type and scale of the data may require database, computing, or server resources not typically available to graduate students. At the University of Iowa, the Digital Studio for Scholarship and Publishing has established itself as a digital humanities center on campus that can support graduate students with data management and preservation, while also providing resources like subscription programs and server space. Some institutions may house staff with these specializations within their university library, but connecting graduate students and departments with the expertise and resources necessary to successfully execute, maintain, and preserve dissertation-level digital projects can help allay concerns about long-term stability for and access to alternative projects. Such connections can also help graduate students make informed decisions about what platforms or programs are best suited for their project, before investing significant amounts of time in a particular technology.

While connecting graduate students with specialized resources is a significant component of successful digital dissertation projects, collaboration across graduate program administrators is also necessary to navigate the unique and atypical dimensions of an alternative dissertation. In an ideal environment, graduate programs could be proactive in establishing procedures for depositing or archiving the various components of an alternative dissertation, like many graduate colleges have done for creative MFA or DMA theses. But starting conversations and opening lines of communication early with various administrative bodies that oversee thesis deposit can help clarify expectations and procedures early in the planning process for a digital dissertation, so student, advisor, and committee can all have consistent expectations around final deliverables.

One of the most important components of an alternative dissertation project is the selection of an advisor and committee to help guide it. On one hand, having faculty with subject area expertise who can guide the student in crafting, developing and articulating the central arguments for the dissertation is necessary and beneficial. However, depending on the institution and department, the faculty with subject area expertise may or may not be familiar with emerging forms of scholarship and the digital methods for analysis and argumentation. A committee comprised of faculty with subject area knowledge and digital scholarship expertise can be a useful way to approach the committee as a collaborative structure who are able to come together to effectively guide and shape the digital dissertation project. The dissertation highlighted in this chapter has co-directors, one with subject-area expertise and one who was a digital scholar in the School of Library and Information Science. Other faculty included individuals with subject area expertise as well as those actively involved in digital and multimodal scholarly communities.

For the student and advisor, one of the most significant challenges of an alternative dissertation project is the continual conversation, negotiation and clarification about how the project is unfolding. A significant thread in that relationship is the process of learning how to ask meaningful questions and provide meaningful progress about a type of scholarship in-process that will likely be unfamiliar to the advisor, and for the student represents their first sustained attempt at a dissertation-scale alternative project. A willingness to clearly define in early stages the core, central, or driving research questions, and how the student will or intends to take up narrative argumentation and digital methods to address those questions, helps the project continue to move forward. Learning to operate on the parallel tracks of 'what the argument is' and 'how the argument is being delivered or communicated' is a starting point for establishing an effective student-advisor relationship for alternative projects. Whether digital or conventional, every dissertation should make a meaningful intervention in ongoing scholarly debates. It is our firm conviction that digital projects can do this, so long as the technologies work in the service of the arguments and not the other way around.

Conclusion

In spite of the many challenges a digital or alternative dissertation project presents, the process of navigating an alternative dissertation can be valuable for students, advisors, and graduate programs. Regardless of any personal reservations faculty may have about digital methods or the 'turn' to digital humanities, the reality remains that graduate students, professional organizations, institutions and employers are becoming increasingly interested and invested in the emerging forms of scholarship facilitated by digital technologies. A proactive approach to developing curricular partnerships, department guidelines, or faculty professional development to facilitate alternative dissertation projects will help current and future students interested in gaining digital skills or undertaking a large-scale alternative project. While many alternative dissertation projects have emerged without necessitating or requiring formal institution or department policy changes, graduate students and faculty will have an easier time navigating the landscape of alternative projects if guidelines and expectations are publicly available and clearly articulated, rather than negotiated in real-time as a graduate student is attempting to propose and craft an alternative dissertation.

For graduate students, a digital dissertation presents the opportunity to construct not only the content and argument of a dissertation, but also to make a variety of choices about how that argument will be produced, represented and communicated. However, the skills necessary to undertake a dissertation-scale digital project are not frequently included in graduate curricula. Thinking concretely about how coursework and other opportunities can be used to become familiar with digital scholarship models and digital methods, while also gaining some level of technical facility, is a valuable step toward evaluating if a digital dissertation project is something worth undertaking. Starting small with a digital project in a graduate seminar or a digital humanities class can help establish confidence and experience with the tools and resources necessary to take on a digital dissertation. Possibly of greatest significance, graduate students interested in pursuing a digital dissertation need to think critically about their long-term research goals and agenda, as well as their personal and professional career aspirations and expectations. Networking and finding mentors within the community of digital scholars working in traditional faculty positions, as well as those working outside traditional faculty roles, is crucial. The continuum of what can constitute 'digital' within a dissertation is broad, and the conversations happening within this collection and across scholarly communities can help students, faculty and programs anticipate and articulate a response to these shifts.

Bibliography

American Studies at the University of Iowa, 'Guidelines for American Studies Graduate Students' (2017), https://clas.uiowa.edu/american-studies/sites/clas.uiowa.edu.american-studies/files/handbook_revised_Fall17.pdf

Carson, A.D., *Owning My Masters: The Rhetorics of Rhymes & Revolutions*, 2016, digital album, https://phd.aydeethegreat.com

Champion, Erik Malcolm, 'DH is Text Heavy, Visualization Light, and Simulation Poor', *Digital Scholarship in the Humanities*, 32 (2017), i25–32, https://doi.org/10.1093/llc/fqw053

Department of History and Art at George Mason University, 'Digital Dissertation Guidelines', https://historyarthistory.gmu.edu/graduate/phd-history/digital-dissertation-guidelines

Humanities Blast, *Digital Humanities Manifesto 2.0.* (2009), http://www.humanitiesblast.com/manifesto/Manifesto_V2.pdf

Jessop, Martyn, 'Digital Visualization as a Scholarly Activity', *Literary and Linguistic Computing*, 23 (2008), 281–93, https://doi.org/10.1093/llc/fqn016

Meeks, Elijah, 'Is Digital Humanities too Text-Heavy?', *Digital Humanities Specialist, Stanford University Libraries* (July 26, 2013), https://dhs.stanford.edu/spatial-humanities/is-digital-humanities-too-text-heavy

Pandey, A. V., et al., 'The Persuasive Power of Data Visualization', *Transactions on Visualization and Computer Graphic, IEEE*, 20 (2014), 2211–20, https://doi.org/10.1109/tvcg.2014.2346419

Sousanis, Nick, *Unflattening* (Cambridge: Harvard University Press, 2015).

Theibault, John, 'Visualizations and Historical arguments', in *Writing History in the Digital Age*, ed. by J. Dougherty and K. Nawrotz (Ann Arbor: University of Michigan Press, 2013), pp. 173–85, https://doi.org/10.2307/j.ctv65sx57.19

Tilton, Lauren, et al., 'Special Issue: Toward a Critically Engaged Digital Practice: American Studies and the Digital Humanities', *American Quarterly*, 70.3 (2018).

Whearty, Bridget, 'Invisible in "The Archive": Librarians, Archivists, and The Caswell Test', *Medieval(ist) Librarians and Archivists: A Roundtable*, 53rd International Congress of Medieval Studies, Kalamazoo, MI, May 10–13, 2018.

8. MADSpace

A Janus-Faced Digital Companion to a PhD Dissertation in Chinese History

Cécile Armand

This chapter is a critical retrospective view of my experience as a PhD candidate in history, whose project made extensive use of digital practices. My argument focuses on MADSpace, a digital platform devoted to a spatial history of advertising in modern Shanghai (1905–49). Born as a digital companion to my dissertation, MADSpace eventually raised new issues and had unexpected effects on my writing process. Beyond my specific case, MADSpace points to the urgent need to establish academic standards for digital scholarship and calls for a better recognition of digital practices by academe.

A Tale of Digital Companionship

Origins of MADSpace

MADSpace was born in 2016 as a digital companion to my PhD project. My research tapped a wide array of primary sources, usually neglected by the existing scholarship in the field: not only newspaper advertisements, but also professional handbooks, business materials, municipal archives (including correspondence, regulations and technical sketches), street photographs and, to a lesser extent, original maps and videos. My primary concern was to create a permanent place to store, organize and connect these multiple sources once digitized or converted into a digital format. Moreover, the spatial approach that I

https://doi.org/10.11647/OBP.0239.08

pursued in my dissertation required specific tools and methods. Indeed, a spatial history of advertising offered an opportunity to experiment with various digital technologies, which in turn could renew the study of advertising, largely dominated by cultural studies to date. Digital methods, I argue, provide new ways of exploring the spatial, social and historical dimensions of advertising. In my research, I relied on a wide range of techniques to produce digital materials aimed to better visualize, analyze and interpret my data. For instance, I used Excel and Fichoz/Actoz (a powerful relational database based on Filemaker) to build databases of advertising agencies and artifacts.[1] I also relied on Geographical Information Systems (GIS) and more basic tools to map the distribution of advertisements in the press and streets[2] of Shanghai, municipal zoning[3] and taxing[4] policies, the networks[5] of advertising agencies and advertising agents' circulations[6] at various scales. I harnessed quantitative analysis tools (Excel) to measure the growth[7] of professional agencies, the relative proportion of Chinese/foreign advertisers,[8] or to measure the rhythmic patterns[9] of advertising spaces. I designed interactive timelines to trace particular series and campaigns[10] or to build a specific periodization[11] for the history of advertising in modern China, embedded in various timescales.[12] In addition, I experimented with intuitive visualizations as alternative ways of displaying my data. For instance, I appropriated mind mapping tools to design various kinds of 'trees' aimed at examining the relationships[13] between multiple actors (companies, municipal authorities, branded goods) and the structure of particular markets (cigarette[14] or health[15]

---

1   See https://madspace.org/cooked/Tables?ID=120
2   See https://madspace.org/cooked/Maps?ID=155
3   See https://madspace.org/cooked/Maps?ID=179
4   See https://madspace.org/cooked/Maps?ID=180
5   See https://madspace.org/cooked/Maps?ID=217
6   See https://madspace.org/cooked/Maps?ID=192
7   See https://madspace.org/cooked/Graphs?ID=316
8   See https://madspace.org/cooked/Graphs?ID=271
9   See https://madspace.org/cooked/Graphs?ID=279
10   See https://madspace.org/cooked/Timelines?ID=106
11   See https://madspace.org/cooked/Timelines?ID=104
12   See https://madspace.org/cooked/Timelines?ID=103
13   See https://madspace.org/cooked/Trees?ID=162
14   See https://madspace.org/cooked/Trees?ID=109
15   See https://madspace.org/cooked/Trees?ID=104

brandscapes). I eventually relied on data sketching methods to create this visuality scale[16] aimed at discussing the assumed 'visual turn' in modern advertising, or these cyclical diagrams aimed at exploring the seasonal effects[17] on commercial images and commodities.

As I was about to start writing my dissertation, MADSpace emerged as a solution to the accumulation of multimedia materials. At the time, I felt the need for a permanent place to store, organize and connect these digitized and born-digital materials, to which I could refer in my dissertation as a mode of quoting primary evidence to support my argument.

A Perfect Match for My PhD Project

MADSpace is hosted by Huma-Num, a 'very large research infrastructure' (*Très Grande Infrastructure de Recherche*—TGIR) supported by the French National Center for Scientific Research (CNRS), with a European and international dimension. Huma-Num provides researchers a variety of tools and services for the processing, dissemination and preservation of digital research data, warranting the long-term sustainability of research projects. Thanks to Huma-Num, scholars do not need to possess the technical skills for developing digital platforms themselves. Therefore, I didn't code myself, but instead, I actively collaborated with the CNRS engineer (Gérald Foliot) who is in charge of developing and maintaining MADSpace among other websites. More concretely, I communicated to him how I conceived of the subdivision into main sections and subsections, the connections between them and the fields contained in each section, and he handled the technical part so as to develop an interface that would match what I needed.

MADSpace is divided into five main sections, which I found the most appropriate way to organize my materials. The *Raw data* section contains my primary sources (archives, printed sources, press advertisements, photos and sketches, original maps and videos). The *Cooked data* section includes the analytical materials produced through digital tools (graphs, maps, trees, timelines and drawings). The *Narratives* section is designed to store the dissertation, research papers and multimedia

---

16   See https://madspace.org/cooked/Drawings?ID=116
17   See https://madspace.org/cooked/Drawings?ID=116

narratives. This section also serves as a research diary including intermediary notes and essays produced during the research process. More conventionally, the *References* section (also in progress) consists of bibliographical references, archival repositories and a bilingual glossary for technical terms. The *Databases* section is made up of four related tables, referring to the four major categories of actors involved in the advertising industry (professional agencies, manufacturing companies, brands and products). The main purpose of this database is to help identify professional actors and to analyze their relationships across space and time. In addition, the horizontal toolbar at the top provides a users' guide that outlines the structure of the platform and specifies its major underlying concepts.

Afterlives

MADSpace has developed far beyond its initial goals. Born as a digital companion to my PhD dissertation, it has eventually become a sustainable platform designed for long-term research projects. In its current state, MADSpace offers three main functions. First, it serves as a digital *repository* aimed at storing, organizing and connecting primary and secondary materials in a cumulative and sustainable way, with a view to make them available for historical research any time. Second, it functions as a digital *laboratory* aimed at experimenting and making transparent every step in the research process, including methodological and technical issues. In contrast to conventional dissertations—that leave only limited room for documenting what I like to call the research *protocol* in order to emphasize the greater proximity between humanities and natural sciences—MADSpace opens a window onto the trial-and-error operations underlying the major findings presented in the final version of the dissertation. Third, it functions as a public *interface* open to sharing and collaboration. As such, it is fully accessible not only to scholars specialized in advertising or modern China, but more generally to anyone interested in urban social history, visual studies or digital practices.

While I initially welcomed MADSpace as a providential solution to the challenge of writing a multimedia-based dissertation, however, this solution in turn raised unexpected dilemmas.

The Traps and Promises of Digital Scholarship
The Curse of Writing

The first challenge was how to connect the platform with a dissertation that remains conventional in its format, based on the page/book model. This is not just a technical issue, but a more fundamental one that questions the very nature of scholarship. How to build a historical narrative directly from digital materials without turning them into mere illustrations? In order to avoid this pitfall, I chose not to include any figures in my dissertation, which would only be text. More exactly, it became a two-faced dissertation, with a digital platform on the one hand, and a rather conventional text, on the other hand, which connects to the platform through a simple system of hyperlinks. Concretely, each hyperlink refers to a unique URL corresponding to a particular piece of evidence that I used to support my argument as a citation. This solution, however, proved far from satisfactory, since it imposed on the reader a constant movement back and forth between the original text and the cited (digital) elements.

The second challenge addressed the possible side effects of digital practices on the writing process. As I was building MADSpace, I developed a penchant toward a 'database' style of writing—to paraphrase a member of my defense committee. Paradoxically, the more materials I accumulated on the platform, the more I tended to expand the core text of my dissertation. While it could have led to a shrinking of the textual content, MADSpace gave birth to a voluminous four-million-character dissertation. The platform opened an infinite space that freed the narrative from the physical constraints of the book format. While the behemoth manuscript may also reflect the lack of time and distance necessary to clean up the final dissertation, it more significantly suggests how digital practices may affect the very crafting of historical narratives.

These two issues eventually merged into this ultimate question: aside from conventional publications, can we design alternate narratives that would fully incorporate our digital experience? Ultimately, MADSpace did not solve the issue of writing dissertations in the digital age. On a practical level, PhD candidates have barely the time or the energy to invest in creating new forms of writing and publishing their findings—a time-consuming and painful task that may appear too risky and would not be rewarded as such. While I truly believe that a PhD project offers the best opportunity for experimenting, my primary concern at the time was more modestly to complete my dissertation so as to obtain my doctoral degree. In order to go beyond conventional or even hybrid dissertations, however, one needs to imagine integrated media that would enable immediate access to digital materials. This is not just an issue of convenience or ergonomics, but more profoundly a way to acknowledge digital practices and their growing part in our research routine.

Beyond my personal story, MADSpace eventually raises the more general challenge of academic recognition for digital experience. How can young scholars in the digital age face the tension between compliance with current standards, and the necessity to cultivate new skills for which there are no defined standards yet? How can we rethink scholarly production in this era of transition in which digital practices receive only a vague and informal recognition, at least in France and Europe? In this respect, MADSpace is just one piece of the ever-growing body of scholarship that points to the urgent need to establish clear standards for better evaluating what these new requirements should be, and how we can integrate them into academic curricula.

Toward Academic Recognition

Promising initiatives like the American Historical Association (AHA)'s guidelines and the HASTAC's conversations, Stanford University Press Digital Projects or Naomi Salmon's Dissertation Form Proposal, have already paved the way. If we are to meet Virginia Kuhn's call for rules that are 'firm enough to ensure rigor yet flexible enough to allow for continued innovation', however, we need to bridge the gap between ideal prescriptions and edifying showcases.[18] New standards must arise from actual realizations. The first step may consist in building a systematic database of digital dissertations, inspired by the HASTAC's cataloging enterprise. This database would not only record the author's name, the title and abstract of the dissertation, the tools and methods used in the research process; it would also include the specific issues

---

18 Virginia Kuhn, 'Embrace and Ambivalence', *Academe*, 99.1 (2013), 8–13, https://eric.ed.gov/?id=EJ1004358

they address, identify their assumptions and the model/s of digital dissertation they suggest. At last, one needs to better categorize and classify the increasing number of digital dissertations that have been produced since Christine Boese's pioneering work,[19] in order to make sense of their variety and their author's creativity. As each dissertation is a unique digital proposal, it is vital to understand their uniqueness before attempting to define a standardized set of rules.

A Digital Laboratory

In order to better grasp the uniqueness of my own digital proposal, I will end by reconnecting my case with other digital dissertations and to broader issues related to digital scholarship. In what sense is MADSpace a 'digital dissertation'? First, it is not primarily a digital publication, or put another way, it is not an *e-dissertation*. It can stand as a PDF alone. It can be can read without any digital device. Some members of my committee even required a printed version. The 'conventional' reader, however, will lose access to the hyperlinks disseminated throughout the main text. While my PDF-dissertation is physically bounded and technically limited, MADSpace is unbounded and potentially unlimited. With the increasing variety of publishing media available to scholars today, the dissertation/book has eventually become only one possible option among many others, as Kuhn cogently put it.[20] My primary concern, however, was not to create a visual argument or to address the issue of fair use, as in Kuhn's case. My dissertation is not so much digital in the mode of writing and publishing, but rather in its particular way of harnessing historical materials in order to make them available and reusable for other stories and for further research. MADSpace is not so much a digital product, but rather a digital process. It is essentially a process because first, it remains open to accumulating and recycling primary and secondary resources. Second, it is sustainable yet flexible enough to evolve and adapt to my changing research needs and interests. Moreover, as it is open to sharing and collaboration with other scholars

---

19   Christine Boese, 'The Ballad of The Internet Nutball: Chaining Rhetorical Visions from the Margins of the Margins to the Mainstream in the Xenaverse' (PhD dissertation, Rensselaer Polytechnic Institute, 1998), http://www.nutball.com/dissertation/

20   Kuhn, 'Embrace and Ambivalence'.

and the more general public, I happen to get feedback from colleagues, genealogists or curious readers outside academia, which encourages me to constantly question my earlier findings and deeper assumptions. In the end, MADSpace is neither an ordinary website nor a digital archive. It is not primarily a digital repository, but rather a digital *laboratory*. As such, MADSpace epitomizes the increasingly experimental and collaborative nature of humanities scholarship in the digital age.

Note on the Terminology

In this chapter, the term *conventional dissertation* (or just *dissertation*) refers to the manuscript I submitted to the defense committee, and *digital dissertation* refers to the print-website complex. This distinction may appear artificial and we must admit that there is no clear-cut dividing line between them. I use these terms for reasons of convenience only. The transitional phase we are experiencing favors hybrid forms of scholarship, and naming things becomes an issue in itself. As objects are changing, we need proper words to designate them. But it is neither my goal nor my ambition here to define what a *conventional* vs. *digital dissertation* is/should be.

Acknowledgements

This research is part of a doctoral project sponsored by the Chiang Ching-Kuo Foundation for International Scholarly Exchange. I am deeply grateful to the Foundation for its generous support for research. As I could not have conducted this digital project alone, I am also grateful to my PhD advisor Christian Henriot (who initially encouraged me to embark on this digital adventure) for his sustained support, to the research engineer Gérald Foliot for his invaluable technical assistance, and to the librarian-engineer Zhang Yu (Lyons for East Asian Studies) for the remarkable design of the platform.

Bibliography

American Historical Association (AHA), 'Guidelines for the Professional Evaluation of Digital Scholarship by Historians', https://www.historians.

org/teaching-and-learning/digital-history-resources/evaluation-of-digital-scholarship-in-history/guidelines-for-the-professional-evaluation-of-digital-scholarship-by-historians

Armand, Cécile, 'Advertising Agencies in Shanghai (1889–1956)' (2018), *MADSpace*, https://madspace.org/cooked/Tables?ID=194

Armand, Cécile, 'Advertising Artifacts in Shanghai Streets (1905–1949)' (2016), *MADSpace*, https://madspace.org/cooked/Tables?ID=120

Armand, Cécile, 'Advertising Professional Circles in Shanghai (1897–1949)' (2016), *MADSpace*, https://madspace.org/cooked/Trees?ID=162

Armand, Cécile, 'Brandscape of Cigarette Advertised in the Shenbao (1934)' (2016), *MADSpace*, https://madspace.org/cooked/Trees?ID=109

Armand, Cécile, 'Brandscape of Medicines Advertised in the Shenbao (1934)' (2016), *MADSpace*, https://madspace.org/cooked/Trees?ID=104

Armand, Cécile, 'Distant Zoning of the Newspaper Shenbao (1914–1949)' (2016), *MADSpace*, https://madspace.org/cooked/Maps?ID=125

Armand, Cécile, 'Heat Maps of Advertising Spaces in Shanghai (1905–1943)' (2016), *MADSpace*, https://madspace.org/cooked/Maps?ID=155

Armand, Cécile, 'F.C. Millington's Biographical Itinerary (1888–1982)' (2016), *MADSpace*, https://madspace.org/cooked/Maps?ID=192

Armand, Cécile, *MADSpace* (2016), https://madspace.org/

Armand, Cécile, 'Multiscalar Timeline of Advertising in Shanghai (1839–1956)' (2016), *MADSpace*, https://madspace.org/cooked/Timelines?ID=103

Armand, Cécile, 'Oriental Advertising Agency's National Network (1924)' (2016), *MADSpace*, https://madspace.org/cooked/Maps?ID=217

Armand, Cécile, 'Periodization of Oriental Advertising Agency's spaces in Shanghai International Settlement (1914)' (2016), *MADSpace*, https://madspace.org/cooked/Timelines?ID=106

Armand, Cécile, 'Polyrhythmic Analysis of Advertising in the Shenbao and North China Daily News (1904–1951)' (2016), *MADSpace*, https://madspace.org/cooked/Graphs?ID=279

Armand, Cécile, 'Population of Advertising Agencies in Modern Shanghai' (2016), *MADSpace*, https://madspace.org/cooked/Graphs?ID=316

Armand, Cécile, 'Professional and Municipal Advertising Geographies in the International Settlement (1931, 1941)' (2016), *MADSpace*, https://madspace.org/cooked/Maps?ID=180

Armand, Cécile, 'Professional and Municipal Advertising Geographies in the French Concession (1926–1943)' (2016), *MADSpace*, https://madspace.org/cooked/Maps?ID=179

Armand, Cécile, 'Scale of Visuality in Newspaper Advertising' (2016), *MADSpace*, https://madspace.org/cooked/Drawings?ID=116

Armand, Cécile, 'Seasonal Effects on Newspaper Advertisements' (2016), *MADSpace*, https://madspace.org/cooked/Drawings?ID=131

Armand, Cécile, 'Sinicization of Advertising Agencies in Shanghai (1905–1956)' (2016), *MADSpace*, https://madspace.org/cooked/Graphs?ID=271

Armand, Cécile, 'User's Guide' (2016), *MADSpace*, https://madspace.org/About/User_Guide

Boese, Christine, 'The Ballad of The Internet Nutball: Chaining Rhetorical Visions from the Margins of the Margins to the Mainstream in the Xenaverse' (PhD dissertation, Rensselaer Polytechnic Institute, 1998), http://www.nutball.com/dissertation/

Centre National de la Recherche Scientifique (CNRS) (French National Center for Scientific Research), *Home page* (Cnrs.fr), http://www.cnrs.fr/index.php/en

Dedieu, Jean-Pierre, *Fichoz*, https://fichoz.hypotheses.org/

Foliot, Gérald, *Foliot.Name*, http://foliot.name/

HASTAC, 'Digital Dissertations Group' (2012), *HASTAC*, https://www.hastac.org/groups/digital-dissertations

HASTAC, 'Workshop: What is a Dissertation? New Models, Methods and Media', *HASTAC*, http://bit.ly/remixthediss-models

Huma-Num, 'About Us', *Huma-Num*, https://www.huma-num.fr/about-us

Kuhn, Virginia, 'Embrace and Ambivalence', *Academe*, 99.1 (2013), 8–13, https://eric.ed.gov/?id=EJ1004358

Salmon, Naomi, 'Dissertation Form Proposal' (2018), *University of Wisconsin Pressbooks*, https://wisc.pb.unizin.org/nsalmondissertationformproposal/

Stanford University Press, 'Digital Projects', *SUP*, https://www.sup.org/digital/

9. Publish Less, Communicate More!

Reflecting the Potentials and Challenges of a Hybrid Self-Publishing Project

Sarah-Mai Dang

In 2014, I finished my doctorate in film studies at Freie Universität Berlin, having writen my thesis on Hollywood chick flicks—conceived through the lens of aesthetic experience, feminist film theory and genre theory. When I was looking for a way to best publish and disseminate my research, the product of more than six years' work, I was surprised to learn that finding an appropriate publisher did not necessarily go hand in hand with disseminating the work in as far-reaching a way as possible. Advice from both senior scholars, as well as colleagues who had already been through the doctoral process, was, first and foremost, to look for a publisher with an outstanding reputation within the disciplinary community. The potential reach seemed to be of secondary importance, the conditions of the publishing contract of no relevance at all.

The fact that authors in the humanities usually receive little if any monetary compensation while at the same time assigning all their rights of use exclusively to the publisher is not a significant issue for most scholars.[1] The publication itself is enough of a reward for many since

---

[1] The rights of use are necessary for publishers in order to carry out marketing measures and produce several versions of a book (paperback, hardcover, e-pub, open access). However, only a minority of the publishers makes use of the various promotion and distribution possibilities. For this reason, in my view, transferring the rights of use constitutes rather a disadvantage for the authors restricting them to freely disseminate their work.

  https://doi.org/10.11647/OBP.0239.09

it is an indication of expertise, a criterion for tenure and for research grants. Seema Rawat and Sanjay Meena (2014) even claim that 'most of the published research works are done just to improve the curriculum vitae (CV)' without actually carrying scholarship forward.[2] They argue that, in order to increase their visiblity as academics and subsequently receive further funding, researchers are forced to create 'publishable research' instead of spending time on significant research or teaching. According to Rawat and Meena, while the number of journals has increased, most publications go uncited due to the lack of appreciation or importance. Even though they speak from the perspective of medical research, their critique also applies to the humanities. The emphasis on publishing takes time away from other fundamental scholarly tasks such as developing a thorough research agenda or an innovative teaching concept—or making scholarly knowledge accessible to a broader public.

I did not know what the pressure to publish-or-perish actually meant until finishing my doctorate thesis. Before, my assumption was that scholarship is geared towards advancing and disseminating knowledge. Today, I am far more aware of the academic system's complexity, its implicit requirements and its power structures, particularly when it comes to the economy of reputation.

For the last several years, I have been exploring the academic publishing ecosystem theoretically and practically. The epistemic conditions of knowledge production in the humanities have become one of my main research areas. After first focusing on open access practices, today as a postdoctoral researcher I scrutinize open scholarship more broadly concerning the political, cultural, technological, economic and legal implications. Since finishing my doctorate I have experimented with various publishing formats and started two blogs about scholarly publishing and open science. The visible overall outcome of my exploration is oabooks.de—a hybrid self-publishing dissertation project.[3]

In this chapter, I reflect upon how the project has developed and consider both the potentials, and challenges, of a digital dissertation.

---

2   Seema Rawat and Sanya Meena, 'Publish or Perish: Where are We Heading?', *Journal of Research in Medical Sciences*, 19.2 (2014), 87–89 (at 88), https://www.ncbi.nlm.nih.gov/pmc/articles/PMC3999612/

3   The website is written in German, as are my original doctorate thesis and initial blog. The newly launched Open Media Studies Blog (https://www.zfmedienwissenschaft.de/online/open-media-studies-blog) on the website of the German journal for media studies *Zeitschrift für Medienwissenschaft* also publishes English posts.

I outline the reasons for my decision to publish my dissertation across four different media formats. In doing so, I elaborate on the specifics of various publishing platforms while also touching upon the core question of how to define legitimate scholarship. I conclude by explaining the most significant findings of my project and what areas need further exploration.

I. One Size Does Not Fit All:
Why Publish Four Different Formats?

In Germany, it is mandatory to make the dissertation accessible to the public in order to complete the doctorate. The supervisors (usually two) have to assess and authorize the manuscript before publishing. Sometimes, but not often, they demand slight changes. It is usually the doctoral candidates who decide on how much editing effort they want to invest in the official publication, which, in the humanities, is still typically a printed monograph. Some scholars 'publish' their thesis via microfiche with the university archive before they officially make it accessible as a book. This is an easy and efficient mode of 'publicaiton' because microfiche does not cost much and helps preserve their work in the long term. However, the main reason for choosing this format lies in the possibility to receive the doctoral award fairly quickly, within a couple of months after the defense. Scholars later officially publish their thesis as a printed book—sometimes after major editing, sometimes with only a little revision. The whole publishing process (finding a suitable publisher, revising the dissertation, applying for funding) can easily take up to two years or more. In Germany, it is common for authors in the humanities to pay an academic publisher for the book production, whether it is a thesis, an anthology or a traditional monograph. Due to the relatively small edition of a scholarly book and hence an estimated small profit, academic publishers calculate so-called printing allowances for the book. These can vary—between 2,500 to 7,000 euro per book (excluding editing)—depending on the status and reputation of the publisher and the author.[4]

---

4   This does not include open access. For an open-access monograph in Germany, an extra fee of between 2,500 and 10,000 euro has to be paid by the author to the publishing house.

The printing allowance has to be paid by the author. If working on an externally-funded research project, the publishing costs are calculated in the project finances. Doctoral candidates, postdoctoral researchers and professors, who are regularly employed by a university, and independent scholars have to apply for third-party funding from foundations or organizations such as the *German Publisher Association* or *Verwertungsgesellschaft Wort* (*VG Wort*), the collecting society of authors, in order to pay the printing allowance. If the application is rejected, scholars have to look for a publisher who can produce the book at relatively low costs, since the printing allowance is quite a considerable amount of money for an individual to pay. For some academic publishers, who accept almost every manuscript, these printing allowances create significant income in addition to the regular sales figures. They are sometimes mockingly called 'subsidy presses' by academics outside the humanities community. However, their publishing program is by no means mediocre or poor. While some publishers follow a publishing strategy which includes a broader range of topics (and audiences), others focus rather on specific areas and target groups. A detailed comparison of the various academic publishers in Germany is beyond the scope of this chapter. In principle, in my experience, the quality of a book in terms of content as well as aesthetics and materiality does not necessarily depend on the amount an author invests financially in the publication or the reputation of a publisher.

As a research assistant at an externally-funded Collaborative Research Center of my university at the time, I would have been able to spend €5,000 on publishing. This would have allowed me to choose a well-known publisher for releasing my dissertation as a book.[5] Yet, the more I explored my options, the more I was unwilling to publish in a system that thrives on the symbolic capital of the book, the restrictions by copyright law and traditional gate-keeping structures. Since in 2014 the publishers I had contacted felt rather reluctant toward open access or any form of self-archiving and were also skeptical regarding a digital dissertation, I decided to disseminate my work myself and make the whole publishing process a research project. My goal was to find a way

---

5   It is important to clarify that also in Germany academic publishers have their own principles that govern manuscript acceptance. While some publishers are known for accepting nearly every submission, others are more selective.

of publishing my thesis that meets what I consider to be three essential requirements: a) accessibility, b) broad distribution, and c) expediency for media studies. Assessing the advantages and disadvantage of already existing multimedia platforms such as Scalar (which offered some impressive possibilities, but was too complex for my needs), text-focused content management websites such as MediaCommons Press (which offered a functional framework, but was ultimately too academic for my desired goal of a broader audience) and ebooks (which provide convenient usage but involve a complicated set-up and lack of standardized file formats and thus compatibility), I realized that one size does not fit all. Therefore, and for the purpose of experiment, I published my dissertation in four different ways, each serving a specific purpose: 1) the original PDF, which emphasizes the institutional part of the doctorate, 2) a website, which speaks to a larger readership, 3) a print-on-demand book, which appeals to book lovers and fulfills standards for common dissemination and 4) a traditional book in English, which opens the discourse to an international community and meets typical professional requirements.

1. An Original PDF Version

Fig. 1 Screenshot of the Freie Universität Berlin repository, by Sarah-Mai Dang (2018), https://refubium.fu-berlin.de/discover?filtertype_0=mycoreId&filter_relational_operator_0=equals&filter_0=FUDISS_thesis_000000101486. CC BY 4.0.

To complete my doctorate, I deposited the original dissertation in the university's repository (see Fig. 1). An institutional repository ensures free access, quick retrieval and sustainable archiving. A PDF can easily be downloaded, marked, forwarded and stored. Scholars do not have to pay any fee for uploading. By making the official version available I also wanted to emphasize that a doctorate thesis represents a preliminary result and not a final product. It is important to remind ourselves that research is a perpetual process and that knowledge is always relative and changes over time. Sometimes this seems to be forgotten, even though one of the humanities scholars' key premises is the relative cultural and social constitution of meaning.

2. A User-Friendly Website

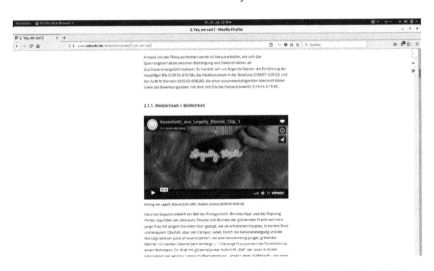

Fig. 2 Screenshot of the second chapter with an embedded film clip, by Sarah-Mai Dang (2018), http://www.oabooks.de/dissertation/web/2-yes-we-can/, CC BY 4.0.

With the help of an editor, I undertook an extensive revision of my thesis before publishing it both as a website with embedded videos and screenshots (see Fig. 2) and a print-on-demand book. The freely available software WordPress allows the author to easily include videos and other media so that the reader can immediately watch the specific film clips while studying the work. When media becomes substantial to the argument, this is very beneficial—if not indispensable—in terms of comprehensibility and transparency for all scholars who analyze audiovisual material, especially film and media scholars. Media objects are for humanities scholars what laboratory measurements are for neuroscientists: research data—the basis of well-founded argumentation. Thus, it is important to make artifacts accessible whenever possible and copyright regulations allow it.[6]

Although the idea of film as an audiovisual language has been a preoccupation of film and media scholars for a long time, the concept of Alexandre Astruc's *caméra stylo* has not really made it into academia as a scholarly practice. Aesthetics and serious reflection still seem to be two distinct dimensions in most researchers' daily routines. Emotional engagement and affective analysis are regarded as (too) subjective and therefore not a legitimate form of scholarship. The concept of an objective, detached researcher is still prevalent in academia and hard to overcome. In my view, however, passion and reason are not incompatible. Nevertheless, due to the easy use of remix technologies and the access to an abundance of material online, more and more scholars have started experimenting with media-rich formats as a site of reflection. Exploring 'new forms of literacy', as Tara McPherson puts it, 'that include authoring and analyzing visual, aural, dynamic, and interactive media', 'multimodal scholars' approach their objects of study differently.[7] They take experience and affect in the context of scholarship seriously.

Catherine Grant, who has initiated various platforms and open-access projects to explore alternative ways of producing and sharing scholarly knowledge, is one of the first media scholars who has been recognized for their videographic essays.[8] Kevin B. Lee is also well-known in the field of audiovisual film studies, even beyond academia.[9]

---

6   For example, the European Court of Justice has ruled that embedding videos is equivalent to including hyperlinks and therefore legal if the copyright holder has already made the film freely accessible online. Since an embedded video is not a copy it does not affect copyright law. See Ilja Braun et al., 'Spielregeln im Internet 1: Durchblicken im Rechte-Dschungel', *Texte 1–8 der Themenreihe zu Rechtsfragen im Netz*, 35 (2017), pp. 38–39, https://irights.info/wp-content/uploads/2017/08/ Spielregeln-im-Internet-Bd-1-2017.pdf. In my case, however, I extracted and uploaded the embedded film clips myself. This is legal because the film scenes function as quotes and not merely embellishment. Unfortunately, there is no fair use doctrine in Germany law, yet.

7   Tara McPherson, 'Introduction: Media Studies and the Digital Humanities', *Cinema Journal*, 48.2 (2009), 119–23 (at 120–21), https://doi.org/10.1353/cj.0.0077

8   See https://vimeo.com/filmstudiesff

9   See https://vimeo.com/kevinblee

Meanwhile, the *European Journal of Media Studies, NECSUS* has created an extra section for audiovisual essays;[10] the Society for Cinema and Media Studies has founded [*in*]*Transition: Journal of Videographic Film & Moving Image Studies*, which publishes video essays exclusively.[11] Scholars, usually with an activist agenda, and artistic researchers, also explore the non-linear participatory potential of interactive web documentaries. The international i-Docs Symposium was convened for the fifth time in Bristol in 2018.[12] Since 2005 the interdisciplinary *Journal of Culture and Technology in a Dynamic Vernacular, Vectors*, has published multimedia texts, which have received multiple awards.[13] The peer-reviewed works are accessible online. Also, the open-source semantic web software Scalar, the template that arose out of *Vectors*, is now widely used by scholars. These are just a few examples of the many projects that film and media studies scholars have initiated in the past years.

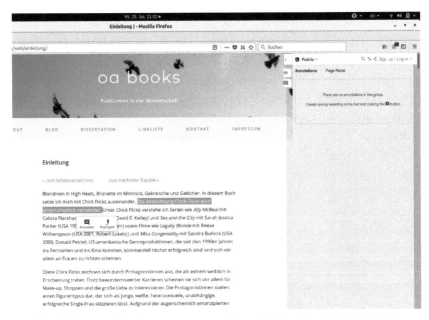

Fig. 3 The screenshot demonstrates annotating with hypothes.is, by Sarah-Mai Dang (2018), http://www.oabooks.de/dissertation/web/einleitung/, CC BY 4.0.

---

10   See https://necsus-ejms.org/
11   See http://mediacommons.org/intransition/
12   See http://i-docs.org/about-interactive-documentary-idocs/i-docs-symposium/
13   See http://vectors.usc.edu/issues/index.php?issue=7

Since I started thinking about how to publish my dissertation *after* it was already written, oabooks.de provides a conventional digitized text rather than a dynamic multimodal website. Yet, in constructing the various versions, I focused on both form and content as well as on the usability of a specific medium. A website can easily be searched, copied and shared. By taking into account essential design elements such as responsivity, navigation, annotation options, typography and color scheme, I optimized the website's accessibility, usability and reach. I chose to build a responsive website which—unlike a PDF—automatically adapts to the size and display of any device and looks good on laptops, tablets and smartphones. Furthermore, I implemented a linked table of contents which directly refers to the chapters as well as footnotes which lead to additional comments. A scroll-to-top button enables the reader to immediately get back to the beginning of the page. I also used hypothes.is, a freely available annotation tool which allows for both collaborative open peer review and personal comments in a private mode (see Fig. 3). In order to provide easy access and a reader-friendly interface I set up an intuitive navigation structure and applied a typography suitable for the web and mobile applications. Of course, the text, which is vital to my project, has to be adjusted to the level of discourse as well. Being trained as well in journalism, I wrote my thesis with a broader public in mind. Similarly, I deliberately went for a vibrant look (by applying pink to the hyperlinks and uploading a non-academic related header) so that the website would not appear too 'serious'. Creating a user-friendly website required lots of consideration of the visual concept and technical aspects as well as sufficient time for designing and programming.

oabooks.de gives access to my dissertation at no cost. All that is needed is a reliable WiFi connection—which is, however, still lacking in most parts of the world.[14] As for the individual scholar, creating a website comes with annual expenses of about $150 for server costs and

---

14 According to a white paper which was published by the World Economic Forum in collaboration with the Boston Consulting Group in 2016, more than half of the world's population do not use the internet due to various reasons such as hard-to-reach areas, lack of basic infrastructure, limited relevant online content, illiteracy, poverty, inequalities (*Internet for All: A Framework for Accelerating Internet Access and Adoption* (Geneva: World Economic Forum, 2016), p. 5, http://www3.weforum.org/docs/WEF_Internet_for_All_Framework_Accelerating_Internet_Access_Adoption_report_2016.pdf).

the domain (about the costs for a few books each year), on top of the hours of labor spent on regular maintenance. To promote the idea of a digital dissertation that differs from a printed form at first I decided not to upload the revised PDF manuscript of the print-on-demand-book. Meanwhile the typeset und formatted manuscript is also available online at various humanities repositories that provide for sustainable archiving, for example the community-led media studies platform MediArXiv.[15] A DOI (digital object identifier) ensures that the digital text can be found, identified and referenced.

3. A Budget Print-On-Demand Book

Fig. 4 Screenshot of my website showing the POD book, by Sarah-Mai Dang (2018), http://www.oabooks.de/dissertation/print-on-demand/, CC BY 4.0.

For various reasons, I chose to also publish the text in a more traditional way. I produced a print-on-demand (POD) book at a relatively low price (€18.90) in addition to the website. Most humanities scholars still prefer a printed version of extensive studies and even shorter articles. A hard copy is still taken to be more legitimate, despite the flaws of the current publishing ecosystem. Even though some traditional publishers

---

15   See https://mediarxiv.org/

are known for printing each manuscript they can make profit with, the book remains a desirable cultural and symbolic capital in academia.

I argue that we need to seriously scrutinize whether a book is really the best form for the inherent processual nature of scholarship. In fact, what Janneke Adema and Gary Hall, among others, call a 'liquid, living' format,[16] such as collaborative wikis, can be much more beneficial. As Adema contends, 'the more "definite" or "final" a text seems (which can be due to language, length, format, style of writing, genre, design, etc.), the harder it becomes for people to engage with it'.[17]

Publications can take a long time. Due to the review process, the author's revision and the publisher's operating procedures, it can take more than two years until a monograph is finally released. In contrast to the natural sciences, research in the humanities might be of less immediate relevance but of longer-term validity. Nevertheless, preprints not only of articles but also of books can facilitate an open, collaborative discussion at a much earlier stage. In this regard, Kathleen Fitzpatrick's book *Planned Obsolescence* might be one of the most well-known examples for more process-oriented than product-oriented book projects,[18] among others such as Jason Mittell's *Complex TV* or Gary Hall's 'Open Book'.[19]

However, a traditional book with an ISBN makes research searchable and accessible through libraries. Libraries play a vital role in terms of dissemination in addition to sustainable archiving. We need to take them into account when discussing openness, accessibility and custodianship. Relatively few people have access to academic libraries though, which is why it is important to make research also freely available online.

I published both the website and the book under the Creative Commons license CC BY-SA 4.0. My work can be shared, copied and used for non-commercial and commercial use if cited properly. The

---

16   'Welcome to the Culture Machine Liquid Books Series Wiki!', http://liquidbooks. pbworks.com/w/page/11135951/FrontPage

17   Janneke Adema, 'A Differential Thesis', *Open Reflections*, ed. by Janneke Adema (July 14, 2015), https://openreflections.wordpress.com/a-differential-thesis/

18   Kathleen Fitzpatrick, *Planned Obsolescence: Publishing, Technology, and the Future of the Academy* (New York: NYU Press, 2011), http://mcpress.media-commons.org/plannedobsolescence/

19   Jason Mittell, *Complex TV: The Poetics of Contemporary Television Storytelling* (New York: NYU Press, 2015), http://mcpress.media-commons.org/complextelevision/ and Gary Hall, 'Open Book', *Media Gifts*, http://www.garyhall.info/open-book/

dissemination must happen under the same license. Many scholars are unsure about the conditions of fair use and copyright. Above all, as authors we bear full responsibilities for our choices. It is important to seek permission to keep the rights of use and negotiate with the publishers. Instead of refraining from exploring new possibilities of digital technologies and sticking to traditional publishing procedures, we should educate ourselves in order to make responsible decisions on how to produce, share and disseminate research.

In order to keep the price of the book as low as possible I did not use any screenshots or other images. These are available online. A paperback of 260 pages costs €18.90 (with a provision of €3.70 to at least partly cover the fees for the editor and graphic designer), the hardcover costs €24.90 (with a provision of €3.10). As of 2016, for the POD production, the Hamburg-based German self-publishing house tredition[20] charged €370. This included thirty-five author copies and the handling of sales and distribution for a minimum of a year.

As with the website, aesthetics were of high relevance in the production process of the book. I wanted the book to be enjoyable not only in terms of content but also of typography and look. By choosing a more striking appearance proposed by a friend who is a designer, the book should encourage a broad audience to engage with the academic work (see Fig. 4). By evoking allusions to cinema seats, hearts and breasts through the same image, the cover references the book's content in a playful way. For me, communicating scholarly knowledge also means showing that passion is an essential factor for acquiring knowledge. The final outcome of the production looks lovely on the bookshelf and raises people's interest.

4. A Traditional Monograph in English

Last but not least, I translated and published my dissertation in English. Instead of spending €5,000 (more than a three-month salary of a part-time working doctoral student at the university) on the printing allowance mentioned above I chose to pay a translator for helping me with an English version of the thesis. With my focus on post-feminism and popular culture, it made sense to share my research with the

---

20   See https://tredition.co.uk/

Fig. 5 Screenshot of the publisher's website offering the English translation of my thesis, by Sarah-Mai Dang (2018), https://www.palgrave.com/de/book/9781137560179, CC BY 4.0.

international community. The role of chick flicks in contemporary media reception has been intensively discussed in the Anglophone world since the late nineties—to a greater degree and for a longer time than in the German-speaking community. Publishing a peer-reviewed book with a prestigious publisher also increases academic reputation and thus hireability (see Fig. 5).

While the concept of peer review is deeply entrenched in many disciplines and academic communities, only in the past few years has peer review—meaning the official practice of an independent external assessment of a work by an authorized expert in the field—become an increasingly important matter in Germany's film and media studies community, with regard to tenure but also as a subject of discussion. To this point, it is usually the colleagues and the editors who help to increase the quality by ensuring that a thesis is based on well-reasoned arguments and supported by evidence. Nowadays, some journals do assess manuscripts via independent peer review, however, most of them do judge texts from an editor's perspective. As for monographs, peer review is still unusual in Germany. The assessment of a book manuscript is carried out by the publisher, respectively an editor of the publishing house.

In my case, several colleagues have read different parts of my dissertation at various stages before the submission. As mentioned above, my supervisors had to authorize the manuscript for publication. I undertook an extensive revision with the help of an editor. In order to enable a post-publication open peer review I implemented the annotation tool hypotheses.is on the website.[21]

The English manuscript, which I wrote and edited with the help of a translator, went through a single-blind peer-review process with one anonymous external reviewer involved, followed by major editing on my part. Since the identity of the reviewer remained unknown to me, it was difficult to engage in a productive dialogue. Nevertheless, the remarks helped to clarify my points, even though I wish I could have directly communicated with the reviewer. To reflect on my experience with traditional publishers in any detail is beyond the scope of this chapter. What I can say, however, is that I have become a strong advocate for reviewing practices which allow for a direct interaction with the 'peers', whether they are authorized by the academic community or simply participating in the debates due to their genuine interests.[22]

II. Why We Do What We Do: Why Scholarly Communication Is an Imperative

The claim for more scholarly communication has led to the expectation that scholars should be able to sell their expertise to the public in an elevator pitch. Yet the duties of an academic job might already feel quite overwhelming due to increasing professional requirements, in particular raising third-party funds. While it is important to open up the university to the wider public and make research more transparent, we need to acknowledge that scholarly communication is a proper professional task that cannot be fulfilled on the side. Like research, teaching and

---

21 Due to the end of service life I decided to uninstall hypotheses.is in 2019. In terms of qualitative feedback, humanities scholars still feel reluctant toward publicly commenting on an article or a book. If it does not involve a popular scholar with a loyal community, comment sections of blogs and open book projects are usually left empty. A visible engagement with my digital dissertation via hypotheses.is did not happen.

22 Kathleen Fitzpatrick, 'Peer Review', in *A New Companion to Digital Humanities*, ed. by Susan Schreibman, Raymond Georg Siemens and John Unsworth (Chichester: Wiley/Blackwell, 2015), pp. 439–48 (at 443–44).

publishing, communication is a fundamental task of scholarship. It is essential when presenting our work at conferences, discussing theories in the classroom—and conveying scholarship publicly. Therefore, it should be valued accordingly for the purpose of job reviews or tenure processes.

While publishing my thesis in four different formats, I have started a blog to reflect upon the project and the social, economic and cultural implications of research more generally. The blog did not come about as a conscious decision for improving scholarly communication but came rather naturally when looking for alternative ways to address a broader audience. A blog allows for a faster, more up-to-date discussion of urgent subjects. It provides a public framework for various formats such as interviews, essays, personal comments, and best-practice reports. Since January 2016, I have posted on topics such as the role of Academia.edu, Elsevier and preprints. Guest authors have also contributed to the blog.

To strengthen the reflection on scholarly practices especially in the German media studies community, in 2018, I started the scholarly interest group Open Media Studies at the German Society for Media Studies (GfM). In addition, I initiated the Open Media Studies blog[23] on the website of the German journal for media studies, *Zeitschrift für Medienwissenschaft*. To speak to a broader readership, the blog also appears at Hypotheses,[24] a multilingual platform for humanities and social science research blogs. Hypotheses is part of the digital publishing infrastructure, OpenEdition. Unlike with my personal blog, with the Open Media Studies blog I try to intervene as little as possible in terms of editing. The co-curated blog is supposed to serve as a collective platform for discussing any subject, position and format related to open access, open scholarship and media studies. Thus, everyone can submit a post at any time. Both my personal and the collective blog, I hope, in addition to my hybrid self-publishing project, help to open up scholarship toward more diverse formats, approaches and practices. We should value scholarship in its wide variety—own initiatives, experiments and projects—and not only as 'the result of a process certified by the usual gatekeepers'.[25] In this sense, I see my endeavor

---

23 See https://www.zfmedienwissenschaft.de/online/open-media-studies-blog

24 See https://mediastudies.hypotheses.org/

25 Steven D. Krause, 'Where Do I List This on My Cv? Considering the Values of Self-published Web Sites. Version 2.0.', *Kairos: A Journal of Rhetoric, Technology, Pedagogy,*

as a best-practice example in which theory and practice are closely intertwined. From this experience, I learned that, in order to effectively intervene in current transitions of digital scholarship, new ideas have to be implemented both from bottom up and top down. Thus, I agree with Virginia Kuhn that 'while we need precedents, no real change will occur without collective action'.[26]

III. Numbers and Stats: Considering the Impact Factor and Drawing a Conclusion

Changing well-established principles is a challenge. Due to the logics of the current academic system, the exploration of unchartered territories can quickly reach its limits. Trying out new forms and formats of publishing is much more of a risk for younger researchers than for senior ones. While doctoral candidates and postdocs might be more open to a digital dissertation, because of their own media practices and daily routines they have to comply with the standard publishing procedures in order to stand a good chance for tenure. Or so they believe. Senior researchers can only gain by experimenting with new publishing options. They have already shown that they can meet the established professional requirements. As a response to the demands made on digital scholars to document and explain their work, Kuhn rightly states, 'explicating one's work is a worthwhile endeavor, but members of review boards still have an ethical imperative to educate *themselves* about the ways in which digital technologies can contribute to rigorous and groundbreaking scholarship'.[27]

Seven years ago, as a doctoral candidate in Germany, who had just finished her thesis, I had to balance between choosing a prestigious publisher—who has built up their reputation by being a quite selective gatekeeper in a decidedly subjective manner of an editor, acting in a rather exclusive way in terms of accepting authors and manuscripts—and reaching the largest audience, by keeping my right of use and

---

12.1 (2007), http://kairos.technorhetoric.net/12.1/binder.html?topoi/krause/index.html

26   Virginia Kuhn, 'Embrace and Ambivalence', *Academe*, 99.1 (2013), https://eric.ed.gov/?id=EJ1004358

27   Ibid.

disseminating research online. These are two difficult goals to combine. Or so I thought. Meanwhile, attitudes have substantially changed towards a more flexible and open access-oriented publishing system. I do not believe that the decision for a hybrid self-publishing project has diminished my chances toward a tenure-track position—if at all calculable. On the contrary, the in-depth analysis of the publishing ecosystem has brought me the status of an expert in the field of open access in the academic community and beyond. Not only have I learned a lot about the publishing business, but I have gained valuable expertise in the application of digital tools and the limitations put in place by copyright law. One of the most significant benefits for me was that I could keep the right of use, which allowed me to experiment with various publishing forms and formats in the first place. In this regard, the results might be viewed as what, according to Adema, Marjorie Perloff has called *differential texts*, '"texts that exist in different material forms, with no single version being the definitive one"'.[28] This approach helps challenging formats taken for granted and highlights the specificities of the various settings. It also shifts the focus from traditional publishing to more diverse forms of scholarly communication.

To create a hybrid self-publishing project was not the plan from the very beginning. It is the product of an extensive two-year study of the academy and the implications of its publishing practices. It is the outcome of lots of editing, programming, designing, translating, discussing and communicating. It is the preliminary result of lots of time-consuming and painstaking work, not to mention the money I have spent on editing, designing and production fees as well as on hosting. While being in a start-up spirit when I began the project, I implemented a few donation options on the website, none of which were used, however. Yet, making money had not been my primary goal, either. Nonetheless, I do see the issue of unpaid digital labor particularly in the platform economy as highly problematic.[29] It is worth noting, however, that scholars also do

---

28   Adema, 'A Differential Thesis'.

29   The issue of free labor for social networking sites of for-profit companies is also of great relevance in the scholarly community considering venture capital-funded platforms such as Academia.edu and ResearchGate. The fact that traditional publishing houses are also profit-led companies which calculate with the unpaid labor of authors, peer reviewers and editors sometimes seems to be forgotten in the discussion of the publishing ecosystem. For a broader discussion see Adema, 'Don't

not receive any financial compensation for editing tasks they perform on a regular basis to set up a manuscript for a publishing house.

In addition to the twenty-five copies I gave out to friends and colleagues for free, two dozen print-on-demand books were sold, which is close to nothing compared to sales numbers of traditional distribution channels. In Germany, academic publishers calculate with about 200–300 copies to be sold in the first two years after the release.

With respect to these numbers, it looks as if the POD project might not have been the most rewarding idea. However, when taking the overall project into account (the website, the translation and the PDF, as well as its wide-ranging effects), the picture looks rather different. The outcome of this endeavor is difficult to assess in terms of numbers, not least because the data is hard to compare. For example, I uploaded the digital version of the revised thesis in German much later, in 2019. Furthermore, since the final manuscript is published under the Creative Commons license CC BY-SA 4.0, it is difficult to track how many times the document has been shared. Also, there is no data available on the original PDF deposited in the university's repository. Moreover, I do not have any data from similar projects. Download statistics do not say anything about the involvement with the text, nor does a page view, or an impact factor. All it does is to show a basic interest in the topic. Nevertheless, the project is clearly not inconsequential. It speaks to an audience that is more diverse than the usual audience reached when publishing via gatekeepers: school teachers, open science activists, scientific journalists, etc. I have received a good deal of feedback from various people from a range of disciplines.

In terms of qualitative feedback, humanities scholars still feel reluctant toward publicly commenting on an article or a book. If it does not involve a popular scholar with a loyal community, comment sections of blogs and open book projects are usually left empty. A visible engagement with my digital dissertation via the annotation program hypothes.is I implemented has also not happened yet. Furthermore, instead of leaving public comments in text boxes, the feedback to the

---

Give Your Labour to Academia.edu: Use It to Strengthen the Academic Commons', in *Open Reflections*, ed. by Janneke Adema (April 7, 2016), https://openreflections. wordpress.com/2016/04/07/dont-give-your-labour-to-academia-edu-use-it-to-strengthen-the-academic-commons/

blog posts, which generate about 50 to 100 page views a year, is rather given on a personal level, face to face or via email.

In September 2019, a second hybrid open-access monograph was published with oa books, *Kommunikationsräume. Einführung in die Semiopragmatik*, a German translation of Roger Odin's well-known film-theory book *Les espaces de communication. Introduction à la sémio-pragmatique*.[30] This translation by my colleagues Guido Kirsten, Magali Trautmann, Philipp Blum and Laura Katharina Mücke is the outcome of a three-year publishing project. Like my dissertation the book is available as PDF and POD. *Kommunikationsräume* was only possible because everyone involved believed in the idea of opening up scholarship to a broader audience. Due to the lack of funding for independent publishing projects, the project was realized without financial compensation of the translators and editors. The French publisher, Presses Universitaires de Grenoble, kindly agreed to grant the translation license at no extra costs. This second book has been one of the most valuable outputs from the film and media studies community.

The initial hybrid self-publishing project was not only about disseminating my research but also constituted a way to make scholars think about their own workflows and practices. My primary goal was to raise awareness of core questions such as what is scholarship for, and how and why do we do it. I do understand if not everyone wants to invest so much time and private money into presenting their research results. Nevertheless, I encourage everyone to further reflect on how to share and disseminate knowledge by experimenting with various publishing forms and formats. Whether or not these digital formats will be acknowledged by the academic community as a legitimate form of scholarly publication, they make us rethink how we have been presenting research so far and how media—in general, not only digital media—shapes the way we work and think. Thus, instead of aiming for more publications, I suggest we take more advantage of the many possibilities opened up by digital technologies and infrastructures. In this sense, I encourage scholars to publish less and communicate more.

---

30 Roger Odin, *Kommunikationsräume. Einführung in die Semiopragmatik*, trans. by Guido Kirsten, Magali Trautmann, Philipp Blum and Laura Katharina Mücke (Berlin: oa books, 2019).

Acknowledgments

The hybrid self-publishing project oa books would not have been possible without the help of my dear friends and confident and trustful colleagues. Many thanks to Vera Rammelmeyer from the graphic design agency *mischen* for the wonderful book cover design and continuous consulting. As for my dissertation I wish to thank Kerstin Beyerlein for editorial support and Landon Little for translation support; the Collaborative Research Center (SfB 626) for the translation grant; my colleagues for their critical feedback; my friends and family for the moral support, the self-publishing house tradition and Palgrave Macmillan for the cooperation in the publishing processes; the Wikimedia fellowship 'Open Knowledge' for the intellectual support. Comments by Feng-Mei Heberer greatly helped to improve an earlier version of this manuscript. I am also thankful to the editor, Virginia Kuhn, for critically reading the manuscript and providing useful suggestions.

Bibliography

Adema, Janneke, 'A Differential Thesis', *Open Reflections*, ed. by Janneke Adema (July 14, 2015), https://openreflections.wordpress.com/a-differential-thesis/

Adema, Janneke, 'Don't Give Your Labour to Academia.edu: Use It to Strengthen the Academic Commons', in *Open Reflections*, ed. by Janneke Adema (April 7, 2016), https://openreflections.wordpress.com/2016/04/07/dont-give-your-labour-to-academia-edu-use-it-to-strengthen-the-academic-commons/

Braun, Ilja, et al., 'Spielregeln im Internet 1: Durchblicken im Rechte-Dschungel', *Texte 1–8 der Themenreihe zu Rechtsfragen im Netz*, 35 (2017), https://irights.info/wp-content/uploads/2017/08/Spielregeln-im-Internet-Bd-1-2017.pdf

Fitzpatrick, Kathleen, 'Peer Review', in *A New Companion to Digital Humanities*, ed. by Susan Schreibman, Raymond Georg Siemens and John Unsworth (Chichester: Wiley/Blackwell, 2015), pp. 439–48.

Fitzpatrick, Kathleen, *Planned Obsolescence: Publishing, Technology, and the Future of the Academy* (New York: NYU Press, 2011), http://mcpress.mediacommons.org/plannedobsolescence/

Gary Hall, 'Open Book', *Media Gifts*, http://www.garyhall.info/open-book/

Krause, Steven D., 'Where Do I List This on My Cv? Considering the Values of Self-published Web Sites. Version 2.0.', *Kairos: A Journal of Rhetoric, Technology, Pedagogy*, 12.1 (2007), http://kairos.technorhetoric.net/12.1/binder.html?topoi/krause/index.html

Kuhn, Virginia, 'Embrace and Ambivalence', *Academe*, 99.1 (2013), 8–13, https://eric.ed.gov/?id=EJ1004358

McPherson, Tara, 'Introduction: Media Studies and the Digital Humanities', *Cinema Journal*, 48.2 (2009), 119–23, https://doi.org/10.1353/cj.0.0077

Odin, Roger, *Kommunikationsräume. Einführung in die Semiopragmatik*, trans. by Guido Kirsten, Magali Trautmann, Philipp Blum and Laura Katharina Mücke (Berlin: oa books, 2019).

Jason Mittell, *Complex TV: The Poetics of Contemporary Television Storytelling* (New York: NYU Press, 2015), http://mcpress.media-commons.org/complextelevision/

Rawat, Seema, and Sanya Meena, 'Publish or Perish: Where are We Heading?', *Journal of Research in Medical Sciences*, 19.2 (2014), 87–89, https://www.ncbi.nlm.nih.gov/pmc/articles/PMC3999612/

'Welcome to the Culture Machine Liquid Books Series Wiki!', http://liquidbooks.pbworks.com/w/page/11135951/FrontPage

World Economic Forum & Boston Consulting Group, *Internet for All: A Framework for Accelerating Internet Access and Adoption* (Geneva: World Economic Forum, 2016), http://www3.weforum.org/docs/WEF_Internet_for_All_Framework_Accelerating_Internet_Access_Adoption_report_2016.pdf

10. #SocialDiss

Transforming the Dissertation into Networked Knowledge Production

Erin Rose Glass

1. Breaking Out of Scholarly Solitude

As anyone knows who has gone through the process, writing a dissertation can be isolating and wearying at times. Attempting to demonstrate mastery of a subject in this particular genre demands long hours of solitary reading and writing with gratification always yet another revision away. Certainly, much can be gained from this arduous process, but it often involves grim and uncertain stretches that make the frequent reports of mental illness in graduate school unsurprising. While some might view these struggles as an important part of scholarly development, for me, they threatened the possibility that I would become a scholar at all. At a certain point, the dissertation had begun to feel like a barrier between myself and the living, and it was no longer clear why I should sacrifice so much for work that promised neither readers nor employment. I soon realized that the only way I could continue my research was if I figured out how to make the dissertation process feel connected to a real and immediate social world. Throwing my insecurities to the wind, I created a plan to use a variety of online tools to draft my dissertation in public view and invite peers and strangers to read and comment along the way. If I was going to spend so much time on a document that is notorious for never being read, then I wanted to see how I might transform the process itself to be its own reward.

https://doi.org/10.11647/OBP.0239.10

Inspired by experiments in social forms of student and scholarly writing that I had only come to learn of in graduate school (and which I will detail later on in this chapter), I created a plan to solicit public review of my dissertation while in the process of writing it. I began the experiment in late February 2017 by posting a draft of the introduction to a public Google Doc and announced its presence on Twitter, Facebook and in a few emails to friends and colleagues over the next few weeks. On the HASTAC blog, an open network that encourages academics to share ideas related to research and teaching, I wrote a short post announcing the experiment as a 'search of an evermore cooperative, influential, and self-directed student public'. Calling the project #SocialDiss after the hashtag that I used to promote it, I asked:

> To what extent can the general public participate in and benefit from the production of a dissertation? How might the private and anxiety-ridden processes of education be transformed into a public good and social joy? Are the imperfect artifacts of learning to be hidden and disposed of as shameful waste, or might they provide fertile soil for the cultivation of a global learning community? Could the form of the dissertation itself blossom into something more vibrant and responsive to today's world in the process?[1]

The draft of the introduction that I posted was far from perfect, perhaps even cringe-worthy at times. But that was the point. I wanted to push against the crippling fear of being judged for imperfect writing and imperfect thoughts. Why should a graduate student writer feel that their work must be perfect in order to participate in writing communities, especially when such perfection is so difficult to achieve for time-poor, working PhD students? Why should forms of student knowledge production focus solely on the final product rather than also encouraging students to use this process to cultivate community and collaborative intellectual practices? To be fair, the isolation I felt may have been exasperated by my personal circumstances: I was writing my dissertation on top of a demanding full-time job three thousand miles away from the academic community I developed in graduate school. But the feeling

---

1   Erin Glass, 'Announcing #SocialDiss: Transforming the Dissertation into Networked Knowledge Production', (March 24, 2017), *HASTAC*, https://www. hastac.org/blogs/erin-glass/2017/03/24/announcing-socialdiss-transforming-dissertation-networked-knowledge

of isolation was not new—I had felt similarly when I still lived close to my graduate institution—nor did it seem uncommon among graduate students. Student writing requirements can feel isolating at any point in one's education if they restrict the student's ability to participate in social life without returning any social nourishment in return.

I carried out #SocialDiss over the course of fifteen months, right up to my defense in April 2018. Though the final form of the dissertation appears much like any traditional, print-oriented dissertation (apart from its afterword and appendix detailing the #SocialDiss process), it was profoundly influenced by the experimental use of different digital writing and networking technologies. These experimental writing practices not only contributed to the intellectual development of the project, they deepened and broadened my scholarly network in important ways and helped make my humanities research more visible (and hopefully more relatable) to my personal networks. In this chapter, I will recount the background, methods, and outcomes of the #SocialDiss project and some of the lessons I believe it offers about the mediating power of writing environments on our intellectual processes. Though #SocialDiss is not necessarily a project I would recommend be repeated to the letter, I think its outcomes suggest that academic writing, especially at the student level, would benefit from digital infrastructure, practices, and incentives that better support forms of in-progress circulation and feedback.

2. Tools for a Student Public

My thinking around the #SocialDiss project was deeply informed by the research I was carrying out for the dissertation itself, which offered a critical history of the adoption of digital technology by universities for humanities research and teaching. In the dissertation, I argued that universities inadvertently taught students to become passive users, or users who are neither capable of understanding how technology mediates their learning activities nor of collectively shaping and governing these technologies according to their needs. I was particularly interested in word processors as a technology whose conventions have become so normalized in academic practice that their influence on our intellectual and social activities in the university is all but invisible.

For at least the past twenty-five years, a nearly singular vision of word processing technology has dominated the tools that humanities scholars and students use to produce their academic writing. What can be difficult to appreciate is that this particular manifestation of writing technology privileges certain ideas about the needs of writing while downplaying others, particularly the social and collaborative possibilities of writing. Features like the skeuomorphic writing interface, copy and paste, and file saving and duplication functionality are so common in our word processors that one might almost consider them as natural components of the writing process itself. These features, however, are anything but natural; as scholars such as Carolyn Handa argue,[2] each of them represents human decisions based on the perceived needs of the writing activity that the program is intended to support as well as the way the programmer views the process and purpose of writing.

The word processor that we have today is largely shaped by business needs to automate tedious aspects of writing through tools like copy and paste features.[3] While these time-saving features may be welcome additions to our writing environments—and in fact were celebrated by many academics when they first began to explore word processing in the 1980s[4]—they have, in some ways, monopolized the imagination of how digital technologies can support, shape and enhance writing processes. As numerous scholars in the field of computers and composition theorized and explored in the 1980s and early 1990s, writing technologies can also be designed to fruitfully support a much broader range of cognitive and social processes than we see supported in Microsoft Word and other similar word processors and in ways that make a significant difference to the final product and the experience of writing itself.[5] However, despite the exciting research and technical development carried out by academics in this area, business-oriented word processors like Microsoft Word became the norm in the academy

---

2   Carolyn Handa, *Computers and Community: Teaching Composition in the Twenty-First Century* (Portsmouth, NH: Boynton/Cook, 1990).

3   Thomas J. Bergin, 'The Origins of Word Processing Software for Personal Computers: 1976–1985', *IEEE Annals of the History of Computing*, 28.4 (2006), 32–47, https://doi.org/10.1109/mahc.2006.76

4   Gail E. Hawisher et al., *Computers and the Teaching of Writing in American Higher Education, 1979–1994: A History* (Norwood, NJ: Aplex Publishing Group, 1996).

5   Ibid.; William Wresch, *The Computer in Composition Instruction: A Writer's Tool* (Urbana: National Council of Teachers of English, 1984).

as opposed to writing technology informed and developed by academic disciplines.

When I arrived at graduate school, I took it for granted—as I believe many academics still do—that writing software was more or less a neutral utility for facilitating the transfer of thoughts from the private mind of the writer to the public page. That does not mean that I was always pleased with whatever writing software I happened to be using or could not imagine a number of improvements. But, all in all, I accepted writing software for what it was along with the belief that—like most forms of digital technology I encountered—it was something made by technologists somewhere far away without any possibility of receiving or caring about my input. I did not imagine that word processors played an influential role in the development of my conception of what it meant to write, think, or produce knowledge. Nor did I imagine that there could be an entirely different form of software production in which the user community (including those who were not technically skilled) might play a role in designing that software. Even if I was told that in fact there were examples and advocates of community-driven software (such as seen in the free software communities), I am not sure I would have been able to imagine what sort of meaningful or intellectual difference academic participation in software design might make. I was largely blind to how—as Johanna Drucker and Patrik Svensson observe—popular technologies used in scholarly production 'imprint their format features on our thinking and predispose us to organize our thoughts and arguments in conformance with their structuring principles—often from the very beginning of a project's conception'.[6]

I may have very well continued ignoring the way word processing software influenced my scholarly practice—academic writing, after all, is hard enough without critically unpacking the tools one uses in the process. However, an unexpected collision between my research interests and experiences pushed me to consider how academic writing tools covertly influence the ways we conceptualize and carry out scholarly work. In my pursuit of looking for diverse critical perspectives on technology within twentieth-century literature, I came across the

---

6   Johanna Drucker and Patrik Svensson, 'The Why and How of Middleware', *Digital Humanities Quarterly*, 10.2 (2016), http://www.digitalhumanities.org/dhq/vol/10/2/000248/000248.html

poet Amiri Baraka's critique of what was the current dominant form of writing technology at the time that he published in 1971:

> A typewriter?—why shd [sic] it only make use of the tips of the fingers as contact points of flowing multi directional creativity. If I invented a word placing machine, an 'expression-scriber', *if you will*, then I would have a kind of instrument into which I could step & sit or sprawl or hang & use not only my fingers to make words express feelings but elbows, feet, head, behind, and all the sounds I wanted, screams, grunts, taps, itches [...][7]

I found Baraka's words both preposterous and brilliant. A writing machine in which one would need to 'sprawl' and 'hang' in order to write seemed ridiculous in comparison to the practical typewriter, but then again, only because I had absorbed what writing machines like the typewriter and computer taught: that writing is strictly a mental, solemn, and private process. Ironically, however, Baraka's vision for a writing process that involved the movement of the whole body in some ways seemed more practical than today's computers given the way computers often cause painful and debilitating back, wrist and neck issues. Baraka's ability to creatively imagine another possibility for such an ordinary-seeming tool helped me understand that every single aspect of writing technology represented a human decision rather than any sort of natural aspect of the writing process.

During this same early period of my graduate education, several courses I attended required students to post reflections on a course blog or learning management platform as a means of extending our classroom discussion in a virtual space. Some courses even went as far as encouraging us to share our final papers with other students for peer feedback. I was also introduced to the exciting experiments in pre-publication open peer review pioneered by scholars like Kathleen Fitzpatrick, McKenzie Wark and Noah Wardrip-Fruin, who used collaborative tools like CommentPress to solicit public review of their scholarly drafts.

And through my own research, I came to learn that educators have been experimenting with forms of virtual and analogue student collaboration since at least the early 1980s, such as detailed and

---

7   Imamu Amiri Baraka, *Raise, Race, Rays, Raze: Essays Since 1965* (New York: Random House, 1971), p. 156.

advocated for by Kenneth Bruffee,[8] William Wresch[9] and Lester Faigley[10] and championed yet again in the first decade of the twenty-first century by scholars such as by George Siemens,[11] Henry Jenkins et al.[12] and Cathy Davidson and David Theo Goldberg.[13] Encountering these recent and older experiments in social forms of scholarly and student writing further opened my eyes to the way writing tools can reinforce assumptions and practices related to writing, or alternately, open up new and generative possibilities.

Despite the decades-long endorsement of these social forms of academic writing—at least at the level of student writing—it was the first time I had been personally exposed to these practices. And so, while I was eager to partake in the intellectual and social benefits their advocates demonstrated, I also found the experience disorienting and ridden with anxiety. I was excited about the prospect of transforming the solitary activity of writing assignments for courses into an opportunity to exchange ideas with peers and develop intellectual community. I was also curious to explore how the experience and reach of academic activity might change when cultivated in networked environments. But my lived experience of engaging in open peer review for course writing fell somewhat far from these hopes. If writing for a single professor caused anxiety, writing for a class full of strangers could cause one to want to quit graduate education altogether.

Part of the problem may have been due to the fact that I felt suddenly rushed into a new rhetorical situation in which I felt pressure not only to perform 'learning' through my writing but to do so with all the likeability, expertise, personality and confidence that seems necessary

---

8　　Kenneth A. Bruffee, 'Collaborative Learning and the Conversation of Mankind"', *College English*, 46.7 (1984), 635–52.

9　　Wresch, *The Computer*.

10　　Lester Faigley, *Fragments of Rationality: Postmodernity and the Subject of Composition* (Pittsburgh: University of Pittsburgh Press, 1992), https://doi.org/10.2307/j.ctt7zwbhf

11　　George Siemens, 'Connectivism: A Learning Theory for the Digital Age', *International Journal of Instructional Technology and Distance Learning*, 2.1 (2005), 3–10.

12　　H. Jenkins et al., *Confronting the Challenges of Participatory Culture: Media Education for the 21st Century. A Report for the MacArthur Foundation* (Cambridge, MA: MIT Press, 2009), https://doi.org/10.7551/mitpress/8435.001.0001

13　　Cathy N. Davidson and David Theo Goldberg, *The Future of Thinking: Learning Institutions in a Digital Age* (Cambridge, MA: MITPress, 2010), https://doi.org/10.7551/mitpress/8601.001.0001

today for speaking publicly on a social media platform. Though the task of extending classroom discussion in a virtual space may have seemed relatively straightforward, it in fact felt remarkably unclear exactly what one should say in such a space and how one should say it. There was no ongoing student public one could quietly observe in order to develop a sense of how to participate as online discussion sites for courses were popped up and then whisked away with the start and close of every term. The design and functionality of the virtual spaces themselves seemed in conflict with the real needs, practices and sensitivities of student writing, adding further roadblocks to developing genuine and continuous engagement and trust with others. And for numerous technical, institutional and social reasons, none of the writing we posted really had the opportunity to develop a real community of readers in the same way that a tweet, a Facebook post or even a Google Doc have.

Nonetheless, despite the imperfection of these experiences, there were still moments in which the value of networked environments for student writing shone through. Reading the writing of other students gave me a glimpse of their intellectual interests that was not as visible in classroom discussion. Having the opportunity to read their course writing provided the groundwork for connecting with them in real life while also enriching my sense of who might read and even be interested in my own academic writing. Baraka's words echoed in my mind. *why shd* student writing depend on technologies that inhibit the cultivation and sustainability of student publics? *why shd* cat memes and food pictures have digital infrastructure designed to enable their extravagant circulation, but the words that students spend thousands of dollars and hundreds of hours learning to produce remain largely unseen? What sort of writing tool might in fact allow a student public to flourish and how might such a public change the way students thought about the purpose and possibility of their writing? And could our writing technologies and practices help address the fears and anxieties generated in social forms of student writing?

These questions might have withered on the vine but I was taking a course that required a proposal for a digital project and so I had the opportunity to develop the ideas in earnest. One thing led to another and I was soon writing a grant proposal with Urban Education graduate student Jennifer Stoops, English Professor Matthew K. Gold and The CUNY Academic Commons development team for a National Endowment for the Humanities Digital Start Up Grant. We were incredibly fortunate to receive the grant and spent two years developing Social Paper, a platform intended to provide a centralized space for students to network writing and feedback across terms, courses, and disciplines with granular privacy settings for every individual paper.

While developing this tool, I became increasingly concerned about the rise of what Shoshana Zuboff calls 'surveillance capitalism', a form of capital accumulation where personal data is collected to 'predict and modify human behavior as a means to produce revenue and market control'.[14] Though I found that many academics shared concerns about the growing power of surveillance capitalism, it seemed that most of us felt unable to reject its tools in our own knowledge making and communication practices. I became ever more curious about the depoliticization of writing technologies within the university and how they had come to be treated as neutral utilities. But *why shd* we blindly accept the writing technologies that we have been handed, I wondered, especially in an institution whose goal is to cultivate a critical understanding of the world, including the technology that enables us to produce and share that understanding? Our lack of critically evaluating the standard tools of academic knowledge production wasn't only keeping us from shaping them to better serve our intellectual needs. It also helped normalize a passive and helpless acceptance of disturbing forms of surveillance and control carried out in technologies used in the academy and beyond. In this context, I came to see Social Paper not just as a tool for supporting the cultivation of student publics, but as a gesture towards the value of community-governed software within the academy. I looked forward to its launch with the hope that I could begin to use Social Paper instead of proprietary tools for all my academic writing to help showcase the value of a participatory approach to our academic writing technologies.

---

14   Shoshana Zuboff, 'Big Other: Surveillance Capitalism and the Prospects of an Information Civilization', *Journal of Information Technology*, 30.1 (2015), 75–89 (at 75), https://doi.org/10.1057/jit.2015.5

3. Building with Imperfect Tools and Imperfect Words

After we launched a beta version of Social Paper in 2016, it became clear that my plan to use it exclusively for producing and sharing academic writing was unrealistic. While I'm proud of the tool we created and its step towards student-driven software, we did not have enough resources to create a tool capable of competing with writing environments created by major digital companies. By the time I had my dissertation drafts ready, I was concerned that my plan to post them exclusively to Social Paper might sabotage my attempt to generate actual engagement with them given some of the flaws in Social Paper's user experience. On one hand, I wanted to enact an example of what student writing on a student-developed writing platform might look like, but on the other hand, I also wanted to explore the possibility of creating community around one's drafting process. Unfortunately, the two desires no longer seemed compatible. I revised my original plan and decided I would post drafts of the dissertation and reflections on the process across a variety of platforms (including Social Paper) in an open, ongoing experiment. What types of engagement—if any—would I receive on different platforms? How would it affect my scholarship and academic experience? And would I regret being so open with the process? I was not sure what to expect.

Of course, given the scarcity of free time in academic life, I didn't expect that anyone would donate their own small scraps of it to engage with my dissertation project. To my surprise, however, many did. In the weeks that followed my original post I received 125 comments on the draft introduction from eleven different individuals ranging from close colleagues and academic friends to individuals I had only briefly connected with over Twitter at prior conferences. In addition, the project spawned multiple backchannel connections and encounters where folks opted to give me feedback over coffee or email or connected me with other scholars who kindly shared their perspective on my research area. Hundreds more clicked on links related to the project and even friends and family members outside of academia (with whom I rarely discussed my research) began to ask me about some of the topics I wrote about. The professional generosity I encountered during these weeks was humbling and kept my spirits afloat when other challenges made the dissertation journey feel almost impossible. And, for the first time in my graduate education, it felt like the hours of labor I privately spent doing research were at last a visible part of my identity. People everywhere, from all areas of my life, were suddenly asking me about my work! As Brandon Walsh, Head of Graduate Programs in the Scholars' Lab at the University of Virginia, recently told me, my dissertation 'was probably read far more than most dissertations'.

The engagement I received, however, was not only encouraging, but also intellectually invaluable. Altogether, the comments I received represented one of the most wide-ranging and in-depth conversations I've ever had about my dissertation topic and were overflowing with information and perspective that simply could not be found in research alone. It was impossible, in fact, to digest the rich set of criticism, related anecdotes, conceptual suggestions, and text recommendations I received even in the first few weeks of #SocialDiss. Commenters offered everything from tips on my choice of language, personal experiences with computers and Usenet in higher education in the early 1980s, their reading notes posted on GitHub on the transformation of science as a pastime to a profession in the nineteenth and twentieth century, the potential relevance of Derrida's notion of 'pro-gram,' and even jokes! One of the authors that I engaged with in the introduction urged me to think more carefully about my use of the cyborg concept while sharing criticisms of his own prior use of the term. There is even a two-part, nearly 1,000-word comment thread debating the difference between 'programming' and 'scripting', with a passionate discussion of the rather obscure Emacs text editor between two commenters who didn't know each other. Reading these comments was exhilarating—it was as if I discovered a secret library of unknown texts all related to my dissertation subject.

My commenters also gently pointed out grammatical errors, logical oversights and places where the clarity of my writing could be improved. And while only a few years before, such exposure would have horrified me, I now found it relieving to see that such imperfections wouldn't cause my community to discount or ignore the arguments I was attempting to develop. While their feedback did contribute to the unwanted realization that I needed to rewrite the introduction entirely, it also provided me with the sense of a real conversation that made this rewriting feel more purposeful. I rewrote many parts of the dissertation with more confidence and ease, with their voices ringing in my head. I was no longer writing to a void, but to a real community of readers whose interests were clear to me.

Over the next few months, I continued with the #SocialDiss experiment, posting drafts and links on Google Docs, Twitter, Facebook, Hypothes.is, Medium, the HASTAC website, the Modern Language Association's Humanities Commons, CommentPress, Academia.edu and my personal website. I often posted a draft in one place, wrote a short post about the draft (including a link to it) on another site (such as HASTAC), and then linked to the post on Facebook and Twitter with short introductions to the draft. Writing blog posts on websites invested in cultivating their community (such as HASTAC and the MLA Commons) drew considerable engagement as these organizations would promote my posts on their homepage and social media accounts. It also helped me practice describing my research in a variety of contexts while continuously asking myself why my research might matter to broader publics.

As I suspected, I found that platform functionalities and platform communities made a big difference on the tone, type, and amount of engagement my drafts and posts received. Drafts posted on Google Docs, for example, were far more likely to receive comments than drafts posted anywhere else, and when I gave readers a choice between Google Docs and some other platform, a majority would choose the former. I continued to receive interesting surprises in Google Docs comments, such as uncannily useful feedback from Estee Beck, a scholar I was previously unaware of (leading to my discovery of her very useful research) as well as occasional formatting or spacing corrections from unknown individuals as a friendly sign of their passing through. On Facebook, friends left deeply personal comments about forms of depression and isolation that accompanied their dissertation writing. A short Twitter essay that summarized a chapter and tagged scholars I cite in the chapter resulted in generative conversations with two of those scholars that continue to this day. The various results of different forms of engagement are too lengthy to fully describe here, but they have provided a very rich set of examples to draw on for making decisions about ongoing digital projects and making the case for community-driven software in academia to various stakeholders.

As an experiment, #SocialDiss was an attempt to see whether it was possible to generate community around student writing processes that have traditionally been private and at times even isolating. What I hoped to show is that networked forms of sharing writing and feedback can generate invaluable intellectual and social experiences when given the right opportunities, technologies and communities of practice. While I think the project has successfully demonstrated this claim, my aim is not to suggest that all students should consider carrying out similar networked writing projects using the broad range of tools and practices that I employed. Rather, I want to use this experiment to point to what I think student writing *could be* if we developed tools, practices, and a culture of sharing that enabled students to share academic writing and feedback as easily as they share other types of content on social media when they feel ready to do so. This is not to say that all student writing should be public—many parts of thinking and learning demand privacy and sheltered spaces. Nor is it to say that we should naively embrace the logic of social media into our academic practices, such as seen with for profit academic platforms such as Academia.edu and ResearchGate. During my search for an 'evermore cooperative student public', there were times I felt I was teetering all too close to what Gary Hall calls the 'uberfication of the university',[15] a dystopian future where academics have to perform sociality with colleagues and others on social media to maintain a good reputation score. These concerns, however, shouldn't cause us to disregard the valuable potential of networks for student writing. Nor should they convince us that the conventional word processor, a tool developed for office automation, is a more natural and neutral choice. It remains to be seen then what tools and what words might help us bring about a genuine student public. I hope that Social Paper and #SocialDiss can help contribute to our collective imagining of its possibility.

---

15   Gary Hall, *The Uberfication of the University* (Minneapolis: University of Minnesota Press, 2016), https://doi.org/10.5749/9781452958439

Bibliography

Baraka, Imamu Amiri, *Raise, Race, Rays, Raze: Essays Since 1965* (New York: Random House, 1971).

Bergin, Thomas J., 'The Origins of Word Processing Software for Personal Computers: 1976–1985', *IEEE Annals of the History of Computing*, 28.4 (2006), 32–47, https://doi.org/10.1109/mahc.2006.76

Bruffee, Kenneth A., 'Collaborative Learning and the Conversation of Mankind''', *College English*, 46.7 (1984), 635–52.

Davidson, Cathy N., and David Theo Goldberg, *The Future of Thinking: Learning Institutions in a Digital Age* (Cambridge, MA: MITPress, 2010), https://doi.org/10.7551/mitpress/8601.001.0001

Drucker, Johanna, and Patrik Svensson, 'The Why and How of Middleware', *Digital Humanities Quarterly*, 10.2 (2016), http://www.digitalhumanities.org/dhq/vol/10/2/000248/000248.html

Faigley, Lester, *Fragments of Rationality: Postmodernity and the Subject of Composition* (Pittsburgh: University of Pittsburgh Press, 1992), https://doi.org/10.2307/j.ctt7zwbhf

Hall, Gary, *The Uberfication of the University* (Minneapolis: University of Minnesota Press, 2016), https://doi.org/10.5749/9781452958439

Handa, Carolyn, *Computers and Community: Teaching Composition in the Twenty-First Century* (Portsmouth, NH: Boynton/Cook, 1990).

Hawisher, Gail E., et al., *Computers and the Teaching of Writing in American Higher Education, 1979–1994: A History* (Norwood, NJ: Aplex Publishing Group, 1996).

Glass, Erin, 'Announcing #SocialDiss: Transforming the Dissertation into Networked Knowledge Production', (March 24, 2017), *HASTAC*, https://www.hastac.org/blogs/erin-glass/2017/03/24/announcing-socialdiss-transforming-dissertation-networked-knowledge

Jenkins, H., et al., *Confronting the Challenges of Participatory Culture: Media Education for the 21st Century. A Report for the MacArthur Foundation* (Cambridge, MA: MIT Press, 2009), https://doi.org/10.7551/mitpress/8435.001.0001

Siemens, George, 'Connectivism: A Learning Theory for the Digital Age', *International Journal of Instructional Technology and Distance Learning*, 2.1 (2005), 3–10.

Wresch, William, *The Computer in Composition Instruction: A Writer's Tool* (Urbana: National Council of Teachers of English, 1984).

Zuboff, Shoshana, 'Big Other: Surveillance Capitalism and the Prospects of an Information Civilization', *Journal of Information Technology*, 30.1 (2015), 75–89, https://doi.org/10.1057/jit.2015.5

11. Highly Available Dissertations
Open Sourcing Humanities Scholarship

Lisa Tagliaferri

The moment an early career scholar decides how to deposit a dissertation often comes during a fraught period of transition. For those who have options, they must weigh the often unquantifiable benefits of open access against the fear of not being able to turn their dissertation into a viable book for publication to earn tenure. Much of this anxiety stems from concerns about university press policies and copyright with respect to digital publication. Currently, there is not much data around open-access dissertations being published by presses. Additionally, scholars without a publication track record may not be mentally prepared for an additional level of scrutiny from the public after rounds of reviews with their committee. Though dissertations are vetted by important stakeholders in the field, they are often not structurally overhauled by professional editors and polished by copyeditors beyond the writer's own departmental resources. This lack of a final quality assurance process during an already stressful time can cause the dissertator to err on the side of caution and embargo their dissertation for several years if they have the option.[1]

Though I had similar concerns during my dissertation process, I deposited my dissertation in 2017 in the open access repository set up by the libraries of my alma mater (CUNY Academic Works) and licensed my work with a Creative Commons license. Because my dissertation was an interdisciplinary digital humanities project that included source

---

1   Each institutional repository will have its own policies; some repositories may allow embargo while others do not.

    https://doi.org/10.11647/OBP.0239.11

code for data visualizations, text mining and a static website, this choice may seem a little more natural, stemming from the fact that the open-source software movement has championed open repositories as a common practice in the tech community for decades. By conceiving of my project as open source, I was also able to think about version control and opportunities for collaboration on a structural level. As I have long been a member of the public university writ large, contributing back to broader publics was an important legacy I wanted to leave following my doctoral training. The licensing was an aspect that I had to think about more significantly. Due to having done a great deal of work in the tech sphere under a Creative Commons license, I was less intimidated than I may have been had I not had that experience. However, I still relied on mentors who had been in the university and publishing space longer than I had, and was fortunate to be able to discuss my options with very knowledgeable people.

That said, open access, open source and Creative Commons licensing is not for everyone. My choices have been grounded in my own particular opportunities and environment, so I will explain the benefits that can be achieved through doing research this way, but believe that every dissertator should come to their own conclusions based on their own experiences, level of comfort and professional goals.

This chapter explores open access dissertations, beginning with contextualizing openness in the humanities and discussing the histories of open source, open access and Creative Commons. The next section will offer guidance for humanities PhD candidates, their committees and institutions. Stakeholders should consider the multiple levels of open access that can be beneficial for dissertations, and they may wish to adopt additional methodologies from open-source software projects. Finally, I will offer my own dissertation as a case study in the interest of sharing data, beginning with an examination of the open access repository where my dissertation was deposited (CUNY Academic Works), and delving into analytics around my dissertation in the interest of data sharing. This chapter engages the ways that increased openness in the humanities can facilitate innovation in the field, and what the benefits and challenges may be from starting a public-facing research career at that vital moment as a scholar: the dissertation defense and deposit.

The Open Humanities

Within the university, there has been a push towards the public humanities recently, including increased engagement with broader communities through digital humanities, and a turn towards openly sharing articles through scholarly repositories in the field. The Modern Language Association (MLA) received several National Endowment for the Humanities (NEH) grants beginning in 2014 to support its open access initiative, Commons Open Repository Exchange (CORE), which is currently in its beta release and open to all humanities fields. The Open Library of Humanities, a non-profit open access publisher for the humanities and social sciences funded by the Mellon Foundation, was launched in 2015 and supports journals including its own multidisciplinary journal. In terms of the dissertation, open access has become integrated into universities' electronic theses and dissertation (ETD) management systems, with an increasing number of institutions, like Duke University and the City University of New York (CUNY), requiring deposit into an open access repository.[2]

Despite the elevated enthusiasm for the public turn of the humanities, it is not a new interest. Humanistic tradition has long participated in open dialogues with the public: dialectic and public disputation during Antiquity transitioned into public scholarly disputation as a form of pedagogy in the Middle Ages.[3] With the movement from humanism into the Renaissance, increased use of the vernacular in both speech and text opened opportunities for marginalized and uneducated communities to participate in broader intellectual participation within a Latinate world. As Jill Cirasella and Polly Thistlethwaite note, the Italian humanist Colluccio Salutati considered that *disputatio* was an essential form of education as it would be 'absurd to talk with oneself between walls and in solitude'.[4] The printing press allowed increased

---

2   Duke University, The Graduate School, 'ETD Availability', *Duke Graduate School*, https://gradschool.duke.edu/academics/theses-and-dissertations/etd-availability; The Graduate Center Library, 'Dissertations and Theses: Deposit Procedure', *Mina Rees Library Research Guides*, https://libguides.gc.cuny.edu/dissertations/deposit-procedure

3   See, for example, Alex J. Novikoff, *The Medieval Culture of Disputation: Pedagogy, Practice, and Performance* (Philadelphia: University of Pennsylvania Press, 2013), https://doi.org/10.9783/9780812208634

4   Jill Cirasella and Polly Thistlethwaite, 'Open Access and the Graduate Author: A Dissertation Anxiety Manual', in *Open Access and the Future of Scholarly*

dissemination of ideas, which would eventually change the course of scholarship through enabling the journal article and monograph as vehicles for knowledge production. In the nineteenth century, Humboldt University led the modern conception of the PhD that would later be adopted at institutions in the United States. From these beginnings, Kyle Courtney and Emily Kilcer explain, 'at the heart of the conferral of the PhD is an affirmation of an individual's substantial, original, and *public* contribution to one's field'.[5]

Open Source, Open Access and Creative Commons

While the humanities have historically tended to keep public engagement and a commitment to openness at its core, more recently software developers and computer scientists have led the way towards open collaboration that has been facilitated through the spread of the internet. Through sharing source code (the lines of code that implement a program), open source promotes public access to programming files, blueprints, and documentation. Open-source software is freely available to use and redistribute, welcomes collaboration by others, and has been integral to the history of computation.

From the 1950s through the early 1970s, most software was produced in collaboration between academia and industry. This software was typically shipped as public-domain software with source code included, which was necessary in order to implement the software on specific machines.[6] However, computer programs were declared to be protected by copyright within the United States in 1974, causing a decline in publicly available source code.[7] With the rise of proprietary software,

---

Communication: Implementation, ed. by Kevin L. Smith and Katherine A. Dickson (Lanham: Rowman and Littlefield, 2017), pp. 203–24 (at 206), https://www.grad. miami.edu/_assets/pdf/open-access-and-the-graduate-author_-a-dissertation-anxiety-manua.pdf

5   Kyle K. Courtney and Emily Kilcer, 'From Apprehension to Comprehension: Addressing Anxieties about Open Access to ETDs', in *Open Access and the Future of Scholarly Communication: Implementation*, ed. by Kevin L. Smith and Katherine A. Dickson (Lanham: Rowman and Littlefield, 2017), pp. 225–44 (at 226, emphasis mine).

6   Erik von Hippel and Georg von Krogh, 'Open Source Software and the "Private-Collective" Innovation Model: Issues for Organization Science', *Organization Science*, 14.2 (2003), 209–23, https://doi.org/10.2139/ssrn.1410789

7   Jan L. Nussbaum, 'Apple Computer, Inc. v. Franklin Computer Corporation Puts the Byte Back into Copyright Protection for Computer Programs', *Golden Gate*

developers began to create free/libre and open-source alternatives beginning in the 1980s. Richard Stallman and Linus Torvalds became central figures among developers working to make non-proprietary alternatives, developing the GNU Project and Linux, respectively.[8] Serving as the foundation of most web servers and the basis of Android smartphones, Linux has tens of millions of users worldwide.

As the internet promoted collaboration and transparency in the computer science field through open source, the sciences at large began to evaluate their mechanisms of exchange in response to networked computers offering the cheap and easy distribution of digital content. Especially with regard to research funded by the public, debate arose around paid models that hinder the dissemination of scientific ideas.[9] The open access movement began in the 1990s, and continues to proliferate throughout scholarship in STEM, the humanities and social sciences. Open access (OA) refers to the free and unrestricted online access to any published research output, with key statements about open access coming from several open access meetings in the early 2000s.[10]

Along with open access comes the question of licensing. For some, having a less restrictive copyright is more consistent with open values. The popular Creative Commons Attribution license took inspiration from computing's GNU General Public License of the Free Software Foundation (GNU GPL). Founded in 2001, Creative Commons was designed for works that are not software, allowing creators to either make their work available in the public domain, or retain their copyright but make the work free for certain uses and conditions.[11] Among the more restrictive Creative Commons licenses are the Non-Commercial and Share-Alike provisions, allowing creators to keep more rights reserved if that is their preference.

University Law Review, 14.2 (1984), 281–308, https://digitalcommons.law.ggu.edu/cgi/viewcontent.cgi?article=1344&context=ggulrev

8   Kathleen Juell, 'A Brief History of Linux' (October 27, 2017), *DigitalOcean*, https://www.digitalocean.com/community/tutorials/brief-history-of-linux

9   Mikael Laakso et al., 'The Development of Open Access Journal Publishing from 1993 to 2009', *PLoS ONE*, 6.6 (2011), e20961, https://doi.org/10.1371/journal.pone.0020961

10 Cold Spring Harbor Laboratory, Library & Archives LibGuides, 'Guide to Open Access: What Is Open Access (OA)?', *CSH*, https://cshl.libguides.com/c.php?g=474046&p=3243847

11 Creative Commons Wiki, 'History' (2011), *Creative Commons*, https://wiki.creativecommons.org/wiki/History

Open source, open access and Creative Commons seek to make work more discoverable and implementable by broader publics. These each can intersect with the choices that are made during a dissertation process (especially a digital dissertation process). Learning more about the histories of each of these frameworks can support the decision making that needs to be done when depositing and disseminating a dissertation.

How To Make a Dissertation Highly Available

With an understanding of open source, open access and Creative Commons licensing, a dissertator can make an informed decision about how best to share their work. In addition to learning about each of the frameworks, they would be well served to have a sense of how each of these is being implemented by scholars, and how open-access works are being received in their given field. Unfortunately, there is not a vast body of quantitative research in this area, but important future work should come through increased interest.

Assuming that a doctoral candidate wants to make their dissertation highly and widely available, they will want to deposit it in an open-access repository if they are able to do so. Many universities in the United States currently have this deposit option available, and a soon-to-be PhD graduate should check in with their institutional library about what their options are in terms of open access. In some cases, open-access repositories may enable a time-bound embargo, preventing public readership of the dissertation until a certain date. In a study of University of Salamanca theses deposits from 2006–2011, researchers found that, across knowledge areas, only the humanities fields had fewer open-access deposits than non-open access.[12]

Opting for an embargo period is especially common in humanities fields because the academic monograph is exceptionally tied to obtaining a long-term job (traditionally in the form of a tenure-track position) or receiving tenure. The issue of the embargo is made more urgent due to the early career precarity for scholars in the current

---

12   Tránsito Ferreras-Fernández et al., 'Providing Open Access to PhD Theses: Visibility and Citation Benefits', *Program: Electronic Library and Information Systems*, 50.4 (2016), 399–416 (at 408). https://doi.org/10.1108/prog-04-2016-0039

climate—there are exceptionally more PhDs produced in the United States than there are full-time academic positions (including tenure-track and non-tenure track roles, as well as term-limited fellowships). For example, in the 2014–15 academic year there were 1,145 new history doctorates awarded, and these recipients could immediately compete for the 572 job advertised in the 2015–16 academic year.[13] With tenure-track positions in research institutions still considered to be the metric of success on the academic job market, candidates find themselves in a desperate situation to err on the side of caution in order to not jeopardize their job prospects in any way.

However, the available research demonstrates that allowing the dissertation to be open access upon deposit does not preclude future publishing opportunities. A recent survey found that when considering an open-access dissertation for a manuscript, university presses were generally receptive: 9.8% indicated that these manuscripts are 'always welcome', 43.9% would consider them on a 'case-by-case basis' and 26.8% would like to see dissertations edited to be 'substantially different' prior to consideration.[14] Cirasella and Thistlethwaite expand on these figures: 'Graduate students might initially be alarmed [...] but it is important to remember that publishers consider *all* manuscripts on a case-by-case basis. Similarly, just about all publishers expect dissertation-based manuscripts to differ significantly from the original dissertation...'[15] Because open-access scholarship is often rendered more visible, an open-access dissertation may prove to be more attractive to presses for a variety of reasons, some of which can be quantified via an open access platform.

---

13 Robert B. Townsend and Emily Swafford, 'Conflicting Signals in the Academic Job Market for History' (January 9, 2017), *Perspectives on History*, https://www.historians.org/publications-and-directories/perspectives-on-history/january-2017/conflicting-signals-in-the-academic-job-market-for-history; Scott Jaschik, 'The Shrinking Humanities Job Market: New Analysis Finds the Number of Doctorates Awarded Keeps Rising, even as Number of Job Openings Drops' (August 28, 2017), *Inside Higher Ed*, https://www.insidehighered.com/news/2017/08/28/more-humanities-phds-are-awarded-job-openings-are-disappearing

14 Marisa L. Ramirez et al., 'Do Open Access Electronic Theses and Dissertations Diminish Publishing Opportunities in the Social Sciences and Humanities? Findings from a 2011 Survey of Academic Publishers', *College & Research Libraries*, 73.4 (2013), 368–80 (374), https://doi.org/10.5860/crl-356

15 Cirasella and Thistlethwaite, 'Open Access and the Graduate Author', p. 206 (their emphasis).

One way to encourage an academic publisher to consider a monograph is through providing them with a common metric of academic success: citations. Allowing a work to be open access 'increases citation rates by 50 percent or more'.[16] Open-access work that is more discoverable allows it to be more readily cited by future researchers, and helps aggregators like Google Scholar count those citations, showing the impact factor of scholarly work. Other data that can support a manuscript proposal include page views, downloads, and other metrics that open-access repositories often make available for authors. As noted in Tránsito Ferreras-Fernández et al.'s study, this is not the case for works that are not open access: 'OA repositories can obtain information on the use (visibility), and on the citation (impact) of doctoral theses, this information cannot be obtained in the case of theses that are not on OA'.[17] While it is necessary to consider that not all academic publishers may be receptive to monograph adaptations of dissertation work, the increased visibility and potential for impact can assist in launching one's career after graduate school.

Though humanities fields still privilege single-author publications as the primary measure for success, an open-access dissertation with no embargo period can help to facilitate collaboration. While some areas of humanities specialization, like the digital humanities, are more receptive to collaborative work, a movement towards more research partnerships, and interdisciplinary ones, can work to advance innovation across the humanities and scholarship at large. Though this open-access outcome is more challenging to measure, the increased visibility can enable opportunities for not only collaborative research projects, but also for traditional humanities alliances: conference panel submissions, invited talks, invitations to edited volumes and more. Allowing the dissertation to be visible upon deposit will provide the early career scholar with the time that is often needed in academia to establish a reputation and foster professional relationships that can support their development.

In addition to deciding whether or not to make a dissertation open access, the author should determine how to license it. Some universities,

---

16   Hillary Corbett, 'Out of the Archives and into the World: ETDs and the Consequences of Openness', in *Open Access and the Future of Scholarly Communication: Implementation*, ed. by Kevin L. Smith and Katherine A. Dickson (Lanham: Rowman and Littlefield, 2017), pp. 187–202 (at 198).

17   Ferreras-Fernández et al., 'Providing Open Access to PhD Theses', 403.

like Duke, require a Creative Commons license,[18] so candidates will need to understand the terms of their university's submission policies. An all rights reserved copyright is a more cautious choice, and should certainly be considered when thinking through future publishing opportunities. However, reserving all rights is generally seen as being against the ethos of open access and will likely prevent healthy dissemination of the text. Using a more flexible Creative Commons license can enable others to share the text to various degrees, encouraging collaboration, idea building, and future research in the specific area of the dissertation. There are several different options for Creative Commons licenses (see https://creativecommons.org/licenses/), and an institutional dissertation librarian can often provide insight on license options. In the end, the candidate should be comfortable with the license they are choosing based on their needs and their expectations for the future lives of their dissertation.

Once an open-access dissertation is available in one repository, one of the best things that a scholar can do to make the work more highly available is through making that publication redundant. Digital media continues to present challenges to archivists, and it is important for submitters to do their own due diligence to make their work redundant and highly available through multiple venues and archival channels to ensure continued access. Disseminating the work widely in different spaces can ensure not only the discoverability of the dissertation (increasing opportunities for downloads, citations, collaborations), but will also ensure that the work persists through multiple digital archive channels. Ferreras-Fernández et al.'s study found that submission across platforms did enable increased access and citations. They write, 'PhD theses disseminated through repositories are benefited through interoperability, which allows their dissemination through multiple portals, sites and search engines, thereby increasing their visibility and making them likely to be cited'.[19] When thinking about increased distribution, the author should consider several spaces: their own personal website and servers, field-specific repositories such as MLA's

---

18   Duke University, The Graduate School, 'ETD Copyright Information', *Duke Graduate School*, https://gradschool.duke.edu/academics/theses-and-dissertations/etd-copy right-information

19   Ferreras-Fernández et al., 'Providing Open Access to PhD Theses', p. 413.

CORE, traditional open-source repositories if applicable, institutional or public OER repositories as appropriate, and submission to sites like archive.org and other free eBook databases. In order to maintain as much control over the work as possible, a PhD candidate should seek out open access scholarly repositories that were developed within universities and built with open-source software, this way those submitting scholarship can even improve the platform in which that scholarship resides. That said, when dissertations become highly available in a redundant way, it becomes more difficult to keep track of the number of engagements that are taking place, whether those are pageviews, downloads or citations, but ultimately redundant copies are the most effective way to ensure the work is openly accessible.

For dissertations that incorporate the work of others, there are some more factors to consider with regard to open access. Long-form translations of writing may need to request permissions from the copyright holder or original publisher prior to making the work open access. Requesting permissions can demonstrate good will even in cases when work is licensed under Creative Commons. In addition to university resources, a candidate may wish to consult with organizations like the American Translators Association for guidance. Art historical dissertations or those that include images and other media files will similarly need to follow guidelines for fair use or fair dealing, and potentially consult the rights holder of any media used. Depending on the period and nature of the work, the proper authority may vary: for manuscripts and early printed books one can consult with the holding archive, for the visual arts including film, a holding museum could be a first step but for living artists or artists with foundations one may decide to make contact via official websites. That said, there are an increasing number of open access and Creative Commons-licensed repositories of media that originate through museums, libraries and organizations like Wikimedia Commons. A candidate can read terms of service and licensing information through websites to ensure that their use case in the dissertation falls within given bounds, and, if so, should cite the origin of incorporated media. In all cases, good judgment should be exercised and citation practices appropriate to scholarly endeavors should be followed.

Dissertations that include data, whether collected or accessed by the doctoral candidate, will also need special consideration. In some cases, the author can cite data that is collected by other sources, but they will have to look into terms of service and sharing policies. When candidates have data collected from human subjects, they will have to ensure that they are in compliance with the Institutional Review Board (IRB) and any other relevant bodies as set out by their university. Doctoral students who find and clean data themselves, through web scraping or other methods, should comply to the data store's terms of service. Candidates can decide whether or not to release the data upon deposit of the dissertation. Making the data publicly available immediately could support others' research and will facilitate the peer review of the dissertation findings. However, if the candidate is planning to do more with that data, there may be good reasons for withholding the full data set until a later date. Data can be released in a software repository, be printed in an appendix, or made available as downloads.

As the digital humanities increase in popularity, more dissertations will include source code. If software is created through collaboration with others, it is necessary to agree on whether or not the software project will be open source and under which license it will be released. The Open Source Initiative includes guidance on different types of licenses and best practices. For those coding alone, an understanding of the open-source ecosystem will help guide the project and its release upon deposit of the dissertation. The software code can live in a Git repository[20] separate from the text of the dissertation, and can also be added to the appendix of the dissertation (within reason), or as zip files as part of the institutional deposit or other disseminations. Where to house a Git repository has become more fraught for humanities researchers as large corporations are increasingly controlling large open-source stores. In addition to housing code on a platform like GitHub, software developers may consider using their own servers to run Git. Wherever the code is housed, once it is released as an open-source project on the internet, there are a number of things for the author or other software

---

20 Git refers to an open-source version-control system that allows for collaborating on computer files. Git repositories are popular for housing code and hosting open-source projects. GitHub, GitLab and SourceForge are popular choices to store software projects and source code.

maintainer to consider. While this is outside of the scope of the current discussion, I provide an overview in my article on 'How to Maintain Open-Source Software Projects'.[21]

An open-access dissertation that is part of a well-maintained repository is well set up to be discovered. However, it is possible for the author to further increase the visibility of the dissertation through discussing it in public digital channels. Traditional social media platforms like Twitter can help to spread the open-access dissertation, as can blogs and even video. A landing website for the project can house all of the various elements, especially if there is source code and other media involved. Institutions may have their own blogging platforms, like CUNY Academic Commons, that can enable either a landing page, or offer the opportunity for multiple posts about the research. Wider, non-institutional networks like HASTAC and Humanities Commons can also serve to share research publicly while establishing connections with other scholars and potential collaborators. The dissertation author should take care to evaluate opportunities against their own comfort levels with wider engagement.

CUNY Academic Works

As a doctoral student of the City University of New York's Graduate Center, I was required to deposit in CUNY Academic Works, an open access repository that holds over 20,000 papers written by the CUNY community (current CUNY faculty, students, and staff may submit their scholarly and creative works to this repository). A service of the CUNY Libraries, CUNY Academic Works collects scholarly papers and provides free public access to these as part of the Library's efforts to advance the mission of CUNY as a public university. To date (February 2021), CUNY Academic Commons' papers have been downloaded approximately 5.5 million times collectively—an average of about 119 downloads per paper.[22] This number is a bit higher than the average number of copies purchased of academic monograph titles in the United

---

21 Lisa Tagliaferri, 'How to Maintain Open-Source Software Projects' (October 6, 2016), *DigitalOcean*, https://www.digitalocean.com/community/tutorials/how-to-maintain-open-source-software-projects

22 CUNY Academic Works, *CUNY Academic Works*, https://academicworks.cuny.edu/

States, which was noted as 83 by Michael Zeoli in 2015 (interestingly, the average number of units sold is the same for both monographs of revised dissertations and those that were new works).[23]

Among the benefits listed on its 'About' page,[24] CUNY Academic Works states that papers that are submitted to the repository will become more discoverable by search engines, be securely hosted on a server, have a dedicated URL for long-term access, and be freely accessible to the public including those who may have limited access to scholarship. This multi-pronged approach to open access taps into the affordances of search engine optimization through metadata and cataloguing efforts, digital archiving for a persistent information store of the research, and a commitment to maintain a perpetually free and public website.

While CUNY Academic Works requires that all Graduate Center dissertations, theses and capstone projects be submitted to the repository, not all are immediately available to read and download, as authors are able to set an embargo period. As noted above, the embargo is intended to allow students to keep their research private prior to publishing the work through an academic press or through journal articles.

In addition to submitting to CUNY Academic Works, I was also required to submit to ProQuest as a doctoral recipient at my institution. Many university ETD management systems in the United States still require some form of submission to ProQuest, an information-content and technology company that was founded in 1938. However, as Gail Clement and Fred Rascoe note, there is a growing number of ProQuest optional or 'NoQuest' institutions (requiring no ProQuest submission), including the University of Michigan, Brown University and Stanford University.[25] While CUNY Academic Works is an open-access initiative unlike ProQuest, the repository uses commercial, proprietary software called Digital Commons produced by Bepress, a commercial software firm that was founded in 1999 and is now owned by the RELX Group. Digital Commons is one of the three software platforms recommended

---

23   Cirasella and Thistlethwaite, 'Open Access and the Graduate Author', p. 208.

24   CUNY Academic Works, 'About', *CUNY Academic Works*, https://academicworks. cuny.edu/about.html

25   Gail P. Clement and Fred Rascoe, 'ETD Management & Publishing in the ProQuest System and the University Repository: A Comparative Analysis', *Journal of Librarianship and Scholarly Communication*, 1.4 (2013), 1–28 (at 5), https://doi. org/10.7710/2162-3309.1074

by Google Scholar for academic repositories to be aggregated by the Google service, and the only one that is not a non-profit (see https://scholar.google.com/intl/en/scholar/inclusion.html).

When I submitted my dissertation, I chose to make the dissertation available without an embargo through both CUNY Academic Works and ProQuest to increase its discoverability and to provide academics more familiar with ProQuest the ability to find it there. Dissertation authors can choose to embargo with neither, either, or both the CUNY Academic Works and ProQuest databases to a greater or lesser amount of time.[26]

An institution's repository and the infrastructure platform that it hooks into are discrete entities, just as a library is a community of people that hooks into an infrastructure of books and other resources. As some institutions have more resources than others, the platform that repositories are located in will vary, and it is the onus of the institution to make choices that will best serve their community based on the resources they have available. In my opinion, an ideal open-access scholarly repository would be one that originated within universities and built with open-source software that could be iteratively improved upon by wider communities. However, I understand that this is not always feasible due to many different challenges and limitations. The fact that CUNY Academic Works leverages the platform of a for-profit company was not clear to me at the time of my submission, and I did not know that the owner of the platform could change. The RELX Group, formerly Reed Elsevier, acquired Bepress in August 2017, shortly after my submission.[27] Although Bepress has leaders with roots in academia, it was not built as an open-source and non-profit organization, leaving it susceptible to acquisition by a large conglomerate. However, through their contract, CUNY ensured that this platform provider did not own the content and the metadata associated with the repository.[28] Ann Hawkins,

---

26 The Graduate Center Library, 'Dissertations and Theses: A Note about Databases and Embargoes', *Mina Rees Library Research Guides*, https://libguides.gc.cuny.edu/dissertations/embargoes

27 Robert Cookson, 'Reed Elsevier to Rename Itself RELX Group' (February 26, 2015), *Financial Times*, https://www.ft.com/content/4be90dbe-bd97-11e4-9d09-00144feab7de; David Bond, 'Relx buys Bepress to Boost Academic Publishing' (August 2, 2017), *Financial Times*, https://www.ft.com/content/c6f6c594-7787-11e7-a3e8-60495fe6ca71

28 Megan Wacha, Scholarly Communications Librarian, Office of Library Services (CUNY), personal communication, August 9, 2018.

Miles Kimball and Maura Ives recommend that universities 'make their ETD policies and information about ETDs available prominently and conveniently on their web sites and in their practices', which I believe will only help PhD students navigate their options.[29] Increased transparency and documentation around dissertation depositing can go a long way towards further empowering graduate students, who are often a marginalized population in the university landscape.

A Dissertation and Its Afterlife

At the moment, my dissertation has not been refactored for publication as a monograph or a series of articles, so I cannot yet detail what that process may look like. I can, however, speak to the process of making this digital humanities project open access with open-source code, and the discoverability aspect of it as a digital dissertation, and extrapolate how this may inform the decision making of others.

Treating the subject of the Italian mystic writer Catherine of Siena and seeking to exert her status as a literary author in her own right through traditional literary analysis paired with digital humanities techniques, 'Lyrical Mysticism: The Writing and Reception of Catherine of Siena' is a comparative literature dissertation that I began working on in 2015 and defended on April 19, 2017. My dissertation work was deposited shortly thereafter, and the full text was made available in the CUNY Academic Works repository on May 17, 2017. I licensed the work under the Creative Commons Attribution-Noncommercial 4.0 License, allowing it to be freely shared and adapted for noncommercial purposes. I spoke to librarians at my institution to decide on a license, and those who are looking to use a Creative Commons license should consider the various options against their hopes for discoverability and opportunities for collaboration.

My dissertation consists of 211 textual pages, as well as a digital component that is housed under the website caterina.io and a GitHub code repository (available at https://github.com/ltagliaferri/dissertation).

---

29 Ann R. Hawkins, Miles A. Kimball and Maura Ives, 'Mandatory Open Access Publishing for Electronic Theses and Dissertations: Ethics and Enthusiasm', *The Journal of Academic Librarianship*, 39.1 (2013), 32–60 (at 38), https://doi.org/10.1016/j.acalib.2012.12.003

The code repository, which is a result of the dissertation being a digital humanities project, consists of several different elements. First, there is the full final text of the dissertation as well as the version-controlled text of the dissertation—for example, if you wanted to see all the revision commits of chapter 1, you can see them via this link: https://github. com/ltagliaferri/dissertation/commits/master/Chapter1.txt. I decided to make revisions publicly visible (as repository commits) in order to show the development of the work over time, as well as provide a record of how feedback was being incorporated. Experimenting with how Git can reveal the evolution of a text could be a fruitful endeavor to show the progression of a text over time. (I am still a bit terrified of providing this level of access to my dissertation.) Next, there are the textual files of the writing of Catherine, Dante and Petrarch, which I used to complete comparative analysis through programming. These files were acquired through web scraping digital versions of primary sources (NB: always read the terms of service prior to web scraping), and I performed some programmatic work to add consistency across the files and bundle files together as needed. Next, there are the actual programming files of the digital humanities work I completed in Python, R and JavaScript. Finally, there is the source code of the website that includes interactive visualizations (these were web recorded by the Graduate Center Library for it to be archived in the repository).

Completing a solo programming project—leveraging data analysis, system administration and web development skills that I developed outside of my PhD program—while also conducting medieval and early modern archival research and completing a traditional dissertation to advance the knowledge of the field was challenging. In retrospect, the care around maintenance and archiving my writing and programmatic work along the way were what suffered the most during this endeavor. Ideally, I would have hosted my dissertation Git repository on my own server (rather than on GitHub's company servers) that could have been archived separately by CUNY librarians, but this would have added an additional level of complexity during an already stressful time. Because of my status as a student and GitHub's generous Student Developer Pack offering, I was able to keep my dissertation work private prior to my defense without having to pay for the privilege, which assuaged my fear of being 'scooped'. Even as someone with significant technical knowledge, I would have benefited immensely from a greater university ecosystem that supports digital humanities projects with teams to help with version control, Git and other digital best practices. For others completing software development projects as part of humanities dissertations, I would encourage seeking out best practices from existing open-source projects. Institutions that are committed to digital humanities projects should consider ways to connect humanities scholars with others in technical fields for support in code reviews, repository stores, testing, and continuous integration and delivery. Increasingly, it is important for universities to hire in-house programmers and network administrators in order to best support the technical research of the university that in industry is usually carried out by full teams.

Because of data that is available to me as the author of my dissertation, I do have a sense of how often it is downloaded and where. CUNY Academic Works provides readership reports through Bepress's Author Dashboard that authors can access through the repository software. Having ready access to data as an author is one of the benefits that is often packaged into open access repositories, allowing you to track downloads and other data points around your dissertation. What is important to note is that each repository where your dissertation is included will have its own data metrics, and you may not have access to all of them. ProQuest is much less transparent in terms of download tracking data than most open access repositories, so you may be unable to keep tabs on every download of your dissertation. It is also challenging to have a sense of how frequently dissertations are downloaded across fields or through various services, so comparative analyses through anecdota are what tend to persist.

To add to the available data, I will share mine here. Based on what I have available from CUNY Academic Works beginning on the date of its deposit on May 17, 2017 until the day of this chapter's revision for publication (February 2021), my dissertation was downloaded over 1,500 times. With ranges between 11 to 62 monthly downloads, there has been no consistent decline. To have an understanding of who is accessing my dissertation more broadly, the text has been downloaded in 90 different countries, and by 180 different organizations (including universities and industry). Dissertation authors may notice that both academic and corporate institutions that are interested in hiring the author for an open role are among those downloading their dissertation.

Although unlike open source, we often think of single-authored dissertations as less collaborative in nature, I think we overlook all the ways that collaboration exists in different degrees and in different directions. As I was working on my dissertation within a traditional humanistic field, the text that I produced was heavily scrutinized by advanced scholars, and I am glad that it was. Librarians at the Graduate Center Library provided considerable support in the deposit of both the text and the digital component of the work, and offered important guidance. The larger Graduate Center community and the broader academic networks of my fields also fostered the advancement of my scholarship in many different ways, and I do not wish to take all of this community nurturing for granted. Still, the development side of my dissertation project did not benefit from code reviews, and I did not have any collaborators to help with the digital manifestation of the work in a hands-on manner. It is possible that I did not fully seek out additional assistance, but a ready framework for this did not exist that I could find in the same manner that I found the considerable documentation and guidance for the more traditional parts of the dissertation. Humanities departments and institutions that encourage digital humanities research should work to support it in meaningful ways, providing resources and direction akin to what they provide for traditional research.

If we return to open-source code development as a framework for open access, there are takeaways from a technology approach that can be applied to an open-access digital dissertation. Transparent and clear structures for writing, review, revision, depositing and caring for the afterlife of the dissertation (whether through making the text redundant, marketing the text, or offering recommendations for filing for infringement) can help to support the dissertation writer as they navigate this process. Open-source code development is often done completely in public over time, while dissertations are often completed in a relatively closed-off manner (apart from conference papers, etc.). The differing practices are a result of the fields, as explicated by the Graduate Center's dissertation research librarian Roxanne Shirazi, who writes, 'In recent years, it would seem that humanities and social science scholars are worried about getting publications out of a dissertation, while STEM folks are increasingly concerned with getting publications into a dissertation'.[30] Having the opportunity to open dialogues with public readership can allow authors to incorporate feedback and iterate on the text for a more vigorous, living book that can be developed over time. Open source has a saying, 'commit early and commit often', speaking to the iterative nature of development projects. What could we, as humanists, gain by engaging early and often?

Bibliography

Anthes, Gary, 'Open Source Software No Longer Optional', *Communications of the ACM* 59.8 (2016), 15–17, https://doi.org/10.1145/2949684

Bond, David, 'Relx buys Bepress to Boost Academic Publishing' (August 2, 2017), *Financial Times*, https://www.ft.com/content/c6f6c594-7787-11e7-a3e8-60495fe6ca71

Budapest Open Access Initiative, 'Read the Budapest Open Access Initiative' (February 14, 2002), http://www.budapestopenaccessinitiative.org/read

Cirasella, Jill, and Polly Thistlethwaite, 'Open Access and the Graduate Author: A Dissertation Anxiety Manual', in *Open Access and the Future of Scholarly Communication: Implementation*, ed. by Kevin L. Smith and Katherine A. Dickson (Lanham: Rowman and Littlefield, 2017), pp. 203–24, https://www.grad.miami.edu/_assets/pdf/open-access-and-the-graduate-author_-a-dissertation-anxiety-manua.pdf

Clement, Gail P., and Fred Rascoe, 'ETD Management & Publishing in the ProQuest System and the University Repository: A Comparative Analysis', *Journal of Librarianship and Scholarly Communication*, 1.4 (2013), 1–28, https://doi.org/10.7710/2162-3309.1074

Cold Spring Harbor Laboratory, Library & Archives LibGuides, 'Guide to Open Access: What Is Open Access (OA)?', *CSH*, https://cshl.libguides.com/c.php?g=474046&p=3243847

Cookson, Robert 'Reed Elsevier to Rename Itself RELX Group' (February 26, 2015), *Financial Times*, https://www.ft.com/content/4be90dbe-bd97-11e4-9d09-00144feab7de

Corbett, Hillary, 'Out of the Archives and into the World: ETDs and the Consequences of Openness', in *Open Access and the Future of Scholarly Communication: Implementation*, ed. by Kevin L. Smith and Katherine A. Dickson (Lanham: Rowman and Littlefield, 2017), pp. 187–202.

---

30 Roxanne Shirazi, 'The Doctoral Dissertation and Scholarly Communication', *College & Research Libraries*, 79.1 (2018), https://crln.acrl.org/index.php/crlnews/article/view/16864/18489

Courtney, Kyle K., and Emily Kilcer, 'From Apprehension to Comprehension: Addressing Anxieties about Open Access to ETDs', in *Open Access and the Future of Scholarly Communication: Implementation*, ed. by Kevin L. Smith and Katherine A. Dickson (Lanham: Rowman and Littlefield, 2017), pp. 225–44.

Creative Commons Wiki, 'History' (2011), *Creative Commons*, https://wiki.creativecommons.org/wiki/History

CUNY Academic Works, *CUNY Academic Works*, https://academicworks.cuny.edu/

CUNY Academic Works, 'About', *CUNY Academic Works*, https://academicworks.cuny.edu/about.html

Duke University, The Graduate School, 'ETD Availability', *Duke Graduate School*, https://gradschool.duke.edu/academics/theses-and-dissertations/etd-availability

Duke University, The Graduate School, 'ETD Copyright Information', *Duke Graduate School*, https://gradschool.duke.edu/academics/theses-and-dissertations/etd-copyright-information

The Graduate Center Library, 'Dissertations and Theses: Deposit Procedure', *Mina Rees Library Research Guides*, https://libguides.gc.cuny.edu/dissertations/deposit-procedure

The Graduate Center Library, 'Dissertations and Theses: A Note about Databases and Embargoes', *Mina Rees Library Research Guides*, https://libguides.gc.cuny.edu/dissertations/embargoes

Hawkins, Ann R., Miles A. Kimball and Maura Ives, 'Mandatory Open Access Publishing for Electronic Theses and Dissertations: Ethics and Enthusiasm', *The Journal of Academic Librarianship*, 39.1 (2013), 32–60, https://doi.org/10.1016/j.acalib.2012.12.003

Jaschik, Scott, 'The Shrinking Humanities Job Market: New Analysis Finds the Number of Doctorates Awarded Keeps Rising, even as Number of Job Openings Drops' (August 28, 2017), *Inside Higher Ed*, https://www.insidehighered.com/news/2017/08/28/more-humanities-phds-are-awarded-job-openings-are-disappearing

Juell, Kathleen, 'A Brief History of Linux' (October 27, 2017), *DigitalOcean*, https://www.digitalocean.com/community/tutorials/brief-history-of-linux

Ferreras-Fernández, Tránsito, et al., 'Providing Open Access to PhD Theses: Visibility and Citation Benefits', *Program: Electronic Library and Information Systems*, 50.4 (2016), 399–416. https://doi.org/10.1108/prog-04-2016-0039

Laakso, Mikael, et al., 'The Development of Open Access Journal Publishing from 1993 to 2009', *PLoS ONE*, 6.6 (2011), e20961, https://doi.org/10.1371/journal.pone.0020961

Novikoff, Alex J., *The Medieval Culture of Disputation: Pedagogy, Practice, and Performance* (Philadelphia: University of Pennsylvania Press, 2013), https://doi.org/10.9783/9780812208634

Nussbaum, Jan L., 'Apple Computer, Inc. v. Franklin Computer Corporation Puts the Byte Back into Copyright Protection for Computer Programs', *Golden Gate University Law Review*, 14.2 (1984), 281–308, https://digitalcommons.law.ggu.edu/cgi/viewcontent.cgi?article=1344&context=ggulrev

Ramirez, Marisa L., et al., 'Do Open Access Electronic Theses and Dissertations Diminish Publishing Opportunities in the Social Sciences and Humanities? Findings from a 2011 Survey of Academic Publishers', *College & Research Libraries*, 73.4 (2013), 368–80, https://doi.org/10.5860/crl-356

Shirazi, Roxanne, 'The Doctoral Dissertation and Scholarly Communication', *College & Research Libraries*, 79.1 (2018), https://crln.acrl.org/index.php/crlnews/article/view/16864/18489

Tagliaferri, Lisa, 'How to Maintain Open-Source Software Projects' (October 6, 2016), *DigitalOcean*, https://www.digitalocean.com/community/tutorials/how-to-maintain-open-source-software-projects

Townsend, Robert B., and Emily Swafford, 'Conflicting Signals in the Academic Job Market for History' (January 9, 2017), *Perspectives on History*, https://www.historians.org/publications-and-directories/perspectives-on-history/january-2017/conflicting-signals-in-the-academic-job-market-for-history

Von Hippel, Erik, and Georg von Krogh, 'Open Source Software and the "Private-Collective" Innovation Model: Issues for Organization Science', *Organization Science*, 14.2 (2003), 209–23, https://doi.org/10.2139/ssrn.1410789

12. The Digital Thesis as a Website

SoftPhD.com, from Graphic Design to Online Tools

Anthony Masure

In France, the official deposit of a PhD thesis is the PDF/A file. However, the submission of an A4 manuscript remains required.[1] The purpose of these requirements is to guarantee the evaluation and archiving of PhD theses. However, these standards pose several problems.

In the field of art or design, the burden of these 'presentation standards' can become counterproductive. Requirements to use Times font body 12, double-spaced, etc., suggest that the PhD in art and design would mainly be 'about' practices (outside the PhD thesis). This opposition between content (idea) and form (depreciated matter) has been criticized since the end of the 1960s by the French philosopher Jacques Derrida.[2] One can thus wonder how the same format can serve radically different purposes, as if the composition, the font, the choice of paper, the arrangement of the blocks, etc., did not necessarily cut a so-called 'external meaning'.[3]

---

1   Emeline Brulé and Anthony Masure, 'Le design de la recherche: normes et déplacements du doctorat en design', *Sciences du Design*, 1 (2015), 58–67, https://doi.org/10.3917/sdd.001.0058

2   Jacques Derrida, *De la grammatologie* (Paris: Minuit, 1967).

3   Anthony Masure, 'À défaut d'esthétique: plaidoyer pour un design graphique des publications de recherche', *Sciences du Design*, 8 (2018), 67–78, https://doi.org/10.3917/sdd.008.0067

    https://doi.org/10.11647/OBP.0239.12

On another level, this gap between academic traditions and contemporary reading and writing practices runs the risk of research avoiding contemporary issues.[4] More and more research is focusing on 'natively' digital materials: videos, websites, interactive installations, video games, data sets, etc. The A4 and PDF formats are not adapted to these subjects, other than through 'fixed' screenshots, to embody the instability and dynamics of 'new media objects'.[5]

In terms of access to knowledge, printed versions of PhD theses can only be consulted in French university libraries. One might think that online PDFs could answer this access problem. PDF (Portable Document Format) is a page description language designed by Adobe in 1992 to preserve the formatting of a document regardless of the program used to read it: it is therefore more an 'intermediate' printable file than a document suitable for reading on screen. PDF, for example, is not responsive (resizing blocks on a phone, etc.). In addition, PDF is poorly indexed by search engines and does not easily allow links to specific sections.

In the United States, theses are mainly accessed via ProQuest Dissertation & Theses Global (PQDT Global), a paid indexing service. Designated as an official offsite repository for the US Library of Congress, PQDT Global emphasizes access to the 'full text' (extracted from PDF's files). This is another way of inducing a certain type of format and excluding divergent experiments.

One can therefore wonder about the persistence of A4 and PDF formats at a time when knowledge is mainly carried out on the Web and with half of the views coming from mobile phones. It seems clear that understanding digital culture and web languages would be positive contributions to a PhD. While HTML was originally created in 1993 to describe and share scientific documents, why do so few (French-language) PhD theses deal with the possibilities of the Web? What could provide a rethinking of the modes of writing and knowledge transmission?

To answer these questions, this article will rely mainly on a specific example. Indeed, it was during the writing of a PhD thesis in aesthetics

---

4   Virginia Kuhn, 'Embrace and Ambivalence', *Academe*, 99.1 (2013), 8–13, https://eric.ed.gov/?id=EJ1004358

5   Lev Manovich, *The Language of New Media* (Cambridge, MA: MIT Press, 2001).

(design) under the supervision of philosopher Pierre-Damien Huyghe at the University of Paris 1 Panthéon-Sorbonne (2008–14) that I was confronted with these issues. Entitled 'Program Design, Ways of Doing Digital', this PhD thesis examines the notion of 'program' (software). The main corpus is a set of philosophical texts, historical events and design projects. The matter of situating design within my thesis arose quite quickly. I decided, in agreement with my PhD supervisor, to carry out a demonstrative design work *on* the thesis. To achieve this, I designed various complementary editorial productions: a graphically designed printed version, an interactive PDF file and a dedicated website (www. softPhD.com). The arguments developed in the PhD thesis are thus replayed by these demonstrative objects.

This article proposes, as a first step, to analyze this work in order to show its goals and extensions. We will see secondly how a few French PhD theses (recently defended or about to be) integrate Web media. Finally, we will ask ourselves more broadly about the role of digital technologies in research practices.

Graphic design, thinking by shaping

In France, the printed version of a PhD thesis is required in most defense committees. This medium is for many people the easiest to read. I had to think about its visual form before considering making another digital version. Entirely written in the proprietary Adobe InDesign software (images are not directly embedded in the file, which reduces its weight), the graphic form of the printed version of my PhD thesis echoes the argumentation. My PhD thesis is thus composed of nine parts, alternating historical and conceptual 'elements'. The different parts of the PhD thesis can be read independently and in any order. The fact that design is not—in my opinion—a 'discipline' allows me to use creativity in writing methods.

The main text is accompanied by thumbnails placed in the margins. These pictures are numbered continuously so that the reader can find them in a larger format at the end of each chapter. Iconography does not have an illustrative role: pictures are sometimes 'indirect' links to the text. It is why they are grouped in autonomous picture boards at the end of each chapter. These blue background picture boards can thus be read separately. They offer another point of view on the concepts. A reminder of the pagination in square brackets allows readers to link the picture books with the current text. The black and blue bichromy unifies the whole PhD thesis. On the level of type, the texts are composed in Mr Eaves Sans—a sans-serif typeface designed by Zuzana Licko in 2010 for the Emigre type foundry—chosen for its legibility and for its humanist forms. The long quotations, indented on the left, are composed in Mrs Eaves XL, a companion serif typeface. As you can see in Figure 1, the alternation between the sans-serif and the serif produces a visual rhythm that allows readers to navigate the text.

Fig. 1 Masure, 'Program Design', 2014, double page with thumbnails. Picture by Anthony Masure, http://www.softphd.com, CC BY-NC-SA.

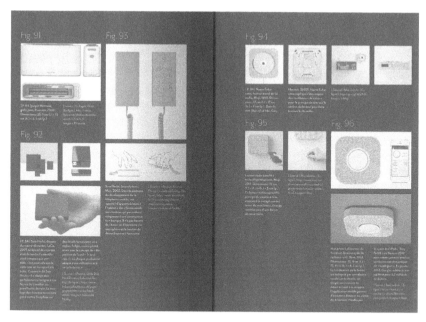

Fig. 2 Masure, 'Program Design', 2014, picture boards. Picture by Anthony Masure, http://www.softphd.com, CC BY-NC-SA.

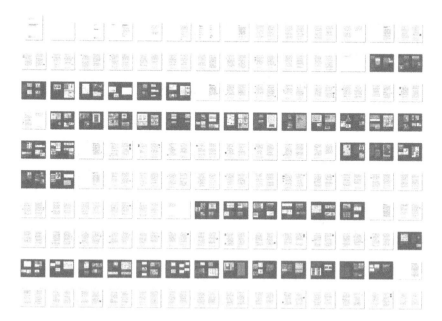

Fig. 3 Masure, 'Program Design', 2014, alternation between current text and picture books. Picture by Anthony Masure, http://www.softphd.com, CC BY-NC-SA.

The layout was finalized by graphic designer Adeline Goyet.[6] It has been designed to provide the reader with a comfortable reading environment (the manuscript contains one million characters). More importantly, the aim was to allow graphic design choices to make sense in regard with the PhD concepts and methodology. The blue background, for example, changes the edge of the PhD thesis into an iconic appearance. The edge points out that the chapters of the thesis are clearly designed as independent. The cross-referencing picture system evokes hypertextual navigation. The table of contents highlights the come and go between historical and conceptual elements, as well as the presence of two appendices (a fiction and a translation). The fiction replays the concepts contended in the thesis in the form of an exhibition project. This 'curatorial fiction' presents a selection of works and projects that connect concepts to tangible initiatives. This method of writing infuses my current research practices.

Fig. 4 Masure, 'Program Design', 2014, edge of the printed thesis. Picture by Anthony Masure, http://www.softphd.com, CC BY-NC-SA.

---

6   Adeline Goyet, *acommeadeline*, http://www.acommeadeline.fr

This non-spectacular layout presents a negotiation with the current standards, some of which could not be redefined: A4 format, number of characters per page, bibliographical notations, white background of the cover, etc. Since design always has to work under constraints, these cannot forbid any attempt at innovation—the most limiting constraints are internalized by researchers in a kind of self-censorship. The current context of academic research in Design, still new in France, allows for a *softness* that will probably be more complicated if things become fixed and stabilized: here, it is essential to maintain a form of flexibility.

SoftPhD.com, Thesis as a Website

The starting point of my PhD thesis questioned conventional forms of reading and multimedia publishing. Thus, it became obvious that I should produce an online version that was as accessible as possible. If I had only produced a printed version, the format would have been in contradiction with my subject matter. The idea of producing a website hosting the full content of the thesis was born of my initiative, after discussion with my thesis supervisor Pierre-Damien Huyghe.[7] We both thought that this idea about media was important, and that it could be generalizable outside of arts and humanities.

As I inquired about what existed in France, I quickly realized that cases were scarce, except for digital extensions of PhD theses (videos, interactive interfaces, etc.) playing as 'supplements' to the main text. In 2014, as far as I know, one of the only PhD theses defended in France to have been fully put online (not in PDF) was that of the philosopher Alexandre Monnin. Its *Philoweb.org* website (2013) has since been deactivated by its author for technical reasons related to the complexities of updates of WordPress and its CommentPress plugin. More generally, if online publication platforms like OpenEdition[8] or Cairn[9] (dedicated to books and articles and not to PhD theses) attend to readability, still their interfaces are difficult to modify. Similarly, self-hosting content management systems (CMS) such as WordPress or Lodel produce

---

7 Anthony Masure, 'Les versions numériques des thèses de doctorat' (November 9, 2017), *Anthony Masure*, http://www.anthonymasure.com/conferences/2017-11-versions-numeriques-theses-doctorat
8 *OpenEdition*, http://books.openedition.org
9 *Cairn*, https://www.cairn-int.info/

'bloated' source code. They favor the use of predetermined templates, without any real creative work.

The unfortunate example of the technical problems at *Philoweb.org* overlaps with one of the arguments in my PhD thesis. Contemporary digital environments are mainly characterized by the industry of coding. This leads to a stack of unintelligible programs. It is therefore necessary to use, when possible, source code that we can master and understand. For these reasons, I chose to design my PhD thesis website without using a CMS. I wanted to establish a relationship between the technical structure and what I contend in the PhD thesis about the readability of the code. In the spirit of openness, all texts were placed under Creative Commons BY-NC-SA license.

This programming work (PHP/CSS/JS) began in July 2014, just after the PDF and printed versions were submitted. I started by 'cleaning' the source code generated from an HTML export from Adobe InDesign. The most laborious aspect was to check all the contents, accents, references, etc. I then built a simple and coherent reading interface with the printed version (same colors, font, etc.), which works at different screen resolutions, including mobile interfaces. A settings menu allows you to adjust the font size and to switch the whole site to a black background. This control of the look and feel is very helpful for persons with visual impairments. Each subsection has its own URL, and can therefore be cited and indexed separately. A section of the website gathers some references collected after the PhD thesis. The jury was able to consult the online version before the defense (November 2014, four months after the submission of the print version), and highlighted its presence as an important element of the research. This website is, in my opinion, the 'true' version of the PhD thesis.

Transfer of Knowledge

This work is not intended to be replicated as it is: each researcher must define what s/he is willing and able to do. For my part, I wanted to renew the PhD thesis reading experience. If we want to move standards that we consider obsolete, then we must try to open other paths. It is a matter of balance: if the design is too experimental, it could fail at legitimizing the method. For this reason, I wrote a foreword that included notes on the

Fig. 5 Masure, website http://www.softPhD.com, 2014, interface variations. Picture by Anthony Masure, http://www.softphd.com, CC BY-NC-SA.

Fig. 6 Masure, website http://www.softPhD.com, 2014, additional resources page. Picture by Anthony Masure, http://www.softphd.com, CC BY-NC-SA.

format for readers who do not necessarily grasp the stakes of graphic design.

The visual form has a very important didactic and is relevant for a teaching role, since it allows us to question how we address each other, how we show ourselves and what we show. The visual homogenization of most PhD theses makes them difficult to read for a broader public, which is in turn counter-productive if we wish to share knowledge. With a website, the presence of pictures and a little interactivity, access becomes easier and accessible to a wider audience. From the beginning, my ambition was to address interface designers and computer developers, an audience not used to reading academic texts. With a dedicated website, I could more easily make my PhD thesis appear in search engines.

In retrospect, this strategy has worked fairly well. The site is consulted approximately twenty times a day and the PDF has been downloaded more than 2000 times since it went online in 2014. This website allows me to easily find fragments, references, etc., for my own work or to send them to someone as a link. Contrary to what one might think, the fact that the PhD thesis is fully available online has not kept publishers from showing interest in my work. Indeed, making theses fully available online can be a distinct advantage, precisely because the feedback we get from readers can enrich the conversation. In this way, I subsequently published the essay *Design et humanités numériques*,[10] part of which is taken from the PhD thesis.

Future of Online Publishing

This work was defended in November 2014. With almost four years of hindsight, beyond the improvement of the consultation interface, a number of improvements seem relevant:

- To further complete the contents of the PhD thesis. In addition to supplementary content (videos, etc.), one could imagine an English translation, which could be achieved in a collaborative way.

---

10   Anthony Masure, *Design et humanités numériques* (Paris: B42, 2017).

- Overcome the opposition between printed format (InDesign) and Web (code editor), which means doing the job twice. This would involve designing a 'web to print' publishing workflow. Web technologies, via CSS Print style sheets, make it possible to define printable layouts. These are now sufficiently strong to be able to compete with traditional layout software. We can think here of free software like Paged.js.[11] This kind of publication workflow would also make it easier to consider other reading formats such as ePub.

- Implement 'versioning' via platforms like GitHub/GitLab. The PhD thesis would be written in a technical environment using Git protocol in order to keep track of the different versions of the text. This method, which comes from software development, is still little used in the context of research content. We can consider here Antoine Fauchié's MA thesis,[12] or *Distill* journal,[13] which use this approach. This method would also have the advantage of allowing the reader to download the sources in .txt/markdown format. The content can thus be easily republished, modified, etc.

- Use the semantic Web and standardize reference notation. This would involve using protocols such as BibTeX, authority databases, etc., so that the contents of the PhD thesis can interact with other datasets.

Other Online PhD Theses Defended in France

Since my defense in 2014, some other cases of online PhD theses have appeared, which are important to mention:

- Saul Pandelakis' film thesis, 'The Hero Who Came Undone: Representation of the Heroic Male Body in American Cinema 1978–2006', defended in 2013, was subsequently put online

---

11   Julien Taquet, 'Behind Paged.js' (July 22, 2018), *PagedMedia.org*, https://www.pagedmedia.org/pagedjs-sneak-peeks/

12   Antoine Fauchié, 'Un mémoire en dépôt' (April 3, 2018), *Quaternum.net*, https://www.quaternum.net/2018/06/04/un-memoire-en-depot

13   *Distill*, https://distill.pub

on the GitBook service,[14] itself based on the GitHub platform. GitBook uses Git protocol to manage different versions of the same document, to facilitate proofreading, etc. Making his PhD thesis available in a browser allows Pandelakis's thesis to be read more easily in fragments (the summary being displayed on the left), at any screen resolution.

Fig. 7 Pandelakis, 'The Hero Who Came Undone', PhD thesis, online version, 2016. Picture by Saul Pandelakis, https://piapandelakis.gitbooks. io/l-heroisme-contrarie/content

- Nolwenn Maudet's PhD thesis on the human-machine interface was put online shortly after its defense.[15] Her PhD thesis was written entirely in HTML in a code editor (Sublime Text). It was then transformed into printable PDF via the HTML2Print tool developed by the designers of the Belgian Open Source Publishing collective.[16] The online version offers

14   Saul Pandelakis, 'L'héroïsme contrarié: formes du corps héroïque masculin dans le cinéma américain 1978–2006', (PhD dissertation, University Paris 3, 2016), https:// piapandelakis.gitbooks.io/l-heroisme-contrarie/content/; *GitBook*, https://www. gitbook.com

15   Nolwenn Maudet, 'Designing Design Tools' (PhD dissertation, University Paris-Saclay, 2017), https://designing-design-tools.nolwennmaudet.com/

16   Open Source Publishing, 'HTML2Print' (2014), http://osp.kitchen/tools/ html2print

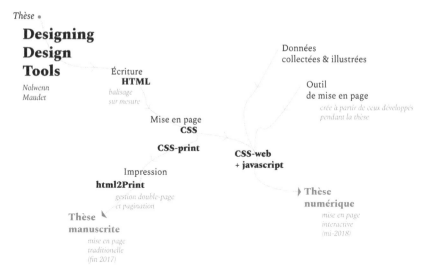

Fig. 8 Maudet, 'Designing Design Tools', 2017, design process. Picture by Nolwenn Maudet, https://designing-design-tools.nolwennmaudet.com/, CC BY-NC-SA.

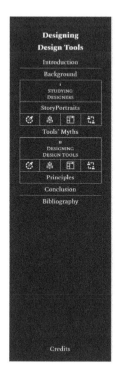

FROM A DESIGNER PERSPECTIVE

Few graphic designers were able to experiment with computers before the era of personal computers. Yet, this limited access did not prevent a few designers to perceive that computing would transform the way they work.

Karl Gerstner - Designing Programs

Figure 14. Karl Gerstner, extract from *Designing Programs*, 1962.

One of the designers who explored the impact of of computing was the swiss Karl Gerstner (**Armstrong, 2009**). In his 1968 book entitled *Designing Programs*, he proposed a manifesto, advocating for a deterministic approach of graphic design. He transposed what he understood from the rigor of computational programs into the typographic grid, turning it into a system (Figure 14). For Gerstner, the computer was mainly a computational machine that he could use to compute all the potential solutions of a design. The design process thus had to be discretized into a set of parameters in order to make it computable. The designer's work was then to cast aside all the bad solutions proposed by the computer and, iteratively, to keep the best one. Even if Gerstner was not using design software at that time, he already had perceived the impact that computing could have on the profession: *"How much computers change – or can change – not only the procedure of the work but the work itself"*. (**Kröplien, 2011**)

Muriel Cooper

If some designers were able to envision how computers could impact their profession, very few had the chance to work with computers before any of the common interface metaphors became ubiquitous. An almost unknown, yet critical example of a designer

Fig. 9 Maudet, 'Designing Design Tools', 2017, design process. Picture by Nolwenn Maudet, https://designing-design-tools.nolwennmaudet.com/, CC BY-NC-SA.

a carefully edited interface, visible in both the work on the navigation menu, type (Spectral, Production Type foundry) and the management of pictures, references, or hypertext links. The texts are released under CC BY-NC-SA license.

- The only other example to date is currently being created by Robin de Mourat, a doctoral student in design. His PhD thesis on academic publication formats (2013–), is currently being prepared at University of Rennes 2 under the supervision of Nicolas Thély (professor in digital art, aesthetics and humanities). Robin de Mourat has developed a writing tool, Peritext, that allows scientific resources to be arranged in a modular and semantic way. His PhD thesis will thus be put online using Peritext.

Peritext

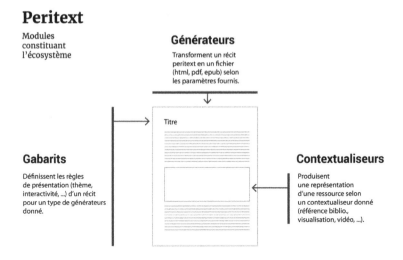

Fig. 10 Robin de Mourat, 'Peritext Program', 2017. Picture by Robin de Mourat, https://peritext.github.io/, CC BY-SA.n

These three examples collectively incorporate many of the improvements I suggested. They show that online PhD theses are far from being constructed using fixed approaches. They take many forms depending on the subject and technological advances.[17] It is striking to note that the

---

17 HASTAC, 'Workshop: What is a Dissertation? New Models, Methods and Media', *HASTAC*, http://bit.ly/remixthediss-models

rare examples of online PhD theses defended in France are the initiative of people claiming to belong to the fields of art and design. We hope that these initiatives will be able to find extensions in other academic disciplines so that the knowledge produced during PhD theses can reach a wider audience. The Web shows that there is a difference between what is published (and which can remain confidential) and what is public (which can be easily read, shared, etc.).

Towards New Frontiers for the PhD

Beyond the transmission and valorization of knowledge, online PhD theses raise the question of PhD theses frontiers and research methodologies. Digital technologies introduce new paradigms that modify the very notion of writing. This begs the question as to why most researchers keep on using Word,[18] even though this proprietary word processing software is not adapted to scientific requirements and contemporary digital culture. With his team, literature researcher Marcello Vitali-Rosati developed the free software Stylo.[19] Stylo is a markdown semantic text editor adapted to the humanities and social sciences to generate both printed and online documents. This kind of approach makes it possible to rethink the place of writing in intellectual activity: the 'tools' of writing are never neutral and strongly engage what is produced.

Online PhD theses also question the boundaries of the PhD. While in France the debates on research-creation are multiplying, they will also have to be helped by reflections on the future of the PhD. The text should remain the central element of knowledge building. But how can the PhD theses also integrate, in addition to pictures, other modes of expression that are not considered simple 'annexes'? How to evaluate PhD theses that modify norms and habits? The evaluation of such practices will undoubtedly require the development of new criteria for assessment and legitimization. While waiting for the latter, doctoral

---

18   Marcello Vitali-Rosati, 'Les chercheurs en SHS savent-ils écrire?' (March 11, 2018), *The Conversation*, https://theconversation.com/les-chercheurs-en-shs-savent-ils-ecrire-93024

19   Marcello Vitali-Rosati, 'Stylo : un éditeur de texte pour les sciences humaines et sociales' (June 3, 2018), *Sens-public.org*, http://blog.sens-public.org/marcello vitalirosati/stylo

students engaging in this type of approach will therefore have to equip themselves with teaching skills and didactics to make the value of their work understood.

The projects analyzed in this chapter show that taking digital culture and web languages into account can help to renew PhD theses. The editorial process of constructing a PhD thesis would therefore benefit from anticipating knowledge transfer. We would then join the initial ambition of the Web, namely a library of scientific documents thought of as a social movement rather than as a technical prowess.[20] The sharing of the computing capacities of machines had made it possible, something unexpected at the time, to gather and transfer knowledge on a global scale. Increasingly threatened by the pressures of capitalism, the defense of this 'free culture' should be the main challenge of academic research.

Bibliography

Berners-Lee, Tim, *Weaving the Web: The Original Design and Ultimate Destiny of the World Wide Web* (New York: Harper Collins, 1999).

Brulé, Emeline, and Anthony Masure, 'Le design de la recherche: normes et déplacements du doctorat en design', *Sciences du Design*, 1 (2015), 58–67, https://doi.org/10.3917/sdd.001.0058

Cairn, https://www.cairn-int.info/

Derrida, Jacques, *De la grammatologie* (Paris: Minuit, 1967).

Distill, https://distill.pub

Fauchié, Antoine, 'Un mémoire en dépôt' (April 3, 2018), *Quaternum.net*, https://www.quaternum.net/2018/06/04/un-memoire-en-depot

GitBook, https://www.gitbook.com

Goyet, Adeline, *acommeadeline*, http://www.acommeadeline.fr

Kuhn, Virginia, 'Embrace and Ambivalence', *Academe*, 99.1 (2013), 8–13, https://eric.ed.gov/?id=EJ1004358

Manovich, Lev, *The Language of New Media* (Cambridge, MA: MIT Press, 2001).

Masure, Anthony, 'À défaut d'esthétique: plaidoyer pour un design graphique des publications de recherche', *Sciences du Design*, 8 (2018), 67–78, https://doi.org/10.3917/sdd.008.0067

---

20   Tim Berners-Lee, *Weaving the Web: The Original Design and Ultimate Destiny of the World Wide Web* (New York: Harper Collins, 1999).

Masure, Anthony, *Design et humanités numériques* (Paris: B42, 2017).

Masure, Anthony, 'Les versions numériques des thèses de doctorat' (November 9, 2017), *Anthony Masure*, http://www.anthonymasure.com/conferences/2017-11-versions-numeriques-theses-doctorat

Maudet, Nolwenn, 'Designing Design Tools' (PhD dissertation, University Paris-Saclay, 2017), https://designing-design-tools.nolwennmaudet.com/

Mourat, Robin de, 'Peritext Program', https://peritext.github.io/

OpenEdition, http://books.openedition.org

Open Source Publishing, 'HTML2Print' (2014), http://osp.kitchen/tools/html2print

Pandelakis, Saul, 'L'héroïsme contrarié: formes du corps héroïque masculin dans le cinéma américain 1978–2006', (PhD dissertation, University Paris 3, 2016), https://piapandelakis.gitbooks.io/l-heroisme-contrarie/content/

HASTAC, 'Workshop: What is a Dissertation? New Models, Methods and Media', *HASTAC*, http://bit.ly/remixthediss-models

Taquet, Julien, 'Behind Paged.js' (July 22, 2018), *PagedMedia.org*, https://www.pagedmedia.org/pagedjs-sneak-peeks/

Vitali-Rosati, Marcello, 'Les chercheurs en SHS savent-ils écrire?' (March 11, 2018), *The Conversation*, https://theconversation.com/les-chercheurs-en-shs-savent-ils-ecrire-93024

Vitali-Rosati, Marcello, 'Stylo : un éditeur de texte pour les sciences humaines et sociales' (June 3, 2018), *Sens-public.org*, http://blog.sens-public.org/marcellovitalirosati/stylo

13. Writing a Dissertation with Images, Sounds and Movements

Cinematic Bricolage

Lena Redman

Backdrop to the Study

In presenting the overview of my doctoral thesis, in this chapter, I focus on *cinematic bricolage* as methodology for the generation, analysis and making meaning of research data. Cinematic bricolage methodology is characterized by the employment of heterogeneous tools and materials that are at hand, as well as by a multimodal system of representing the development of meaning.

In this chapter, I outline the methodology of cinematic bricolage with the use of alphabetic text and infographics. My point here is not to demonstrate my skills as a graphic artist that perhaps give me some advantage in presenting knowledge in an attractive way, and which might be deemed as nonessential. The infographics that I use here are not created to illustrate what is already articulated with words. Their presence here is my argument about the thought embodied in a particular form also being shaped by this form. The infographics—or any other digital mode(s) of expression chosen by the individual—and alphabetic text are two concentric rings of the same ripple, so to speak, that push at and are pushed by each other, resulting in the production of meaning through a continuous expansion of thought.

 https://doi.org/10.11647/OBP.0239.13

Fig. 1 *Writing with infographics and alphabetic text as the concentric rings of a ripple,* Lena Redman (August, 2019).

In this view, this chapter can be characterized as *writing with infographics* and *alphabetic text*.

The correlation between the syntagmatic and paradigmatic dimensions in the production of meaning within Michael Halliday's framework of *systemic-functional linguistics* (*SFL*), served as a departure platform for the study.[1] The paradigmatic dimension is conceptualized as a system of signs and signifiers.

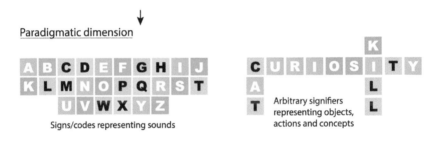

Fig. 2 *Paradigmatic organization of linguistic signs,* Lena Redman (October, 2020).

The paradigmatic dimension is a set of variables that is composed in 'systemic patterns of choice'—syntagmatic structure, produces meaning.[2] The encrypted interplay between the paradigmatic and syntagmatic dimensions constitutes a semiotic system which allows the content of mind to be transferred into an embodied form.

---

1   Michael A. K. Halliday, *Halliday's Introduction to Functional Grammar* (New York: Taylor & Francis Group, 2014).
2   Ibid., loc. 897.

Fig. 3 *Syntagmatic organization of linguistic signs,* Lena Redman (October, 2020).

'Any set of alternatives, together with its conditions of entry, constitutes a system in this technical sense', Halliday maintains.[3] In translating the semiotic system of language as conceptualized by Halliday into a semiotics of digital multimodality, the paradigmatic 'set of alternatives' is associated with database elements.

Fig. 4 *Paradigmatic dimension of multimodal representation: database,*
Lena Redman (October, 2020).

'Syntagmatic ordering in language'—that is, 'conditions of entry' in Halliday's words—in digital multimodality is correlated with the cultural phenomenon of *deep remixability*.[4] This is a 'condition in which everything (not just the content of different media but also

---

3   Ibid.
4   Ibid.

languages, techniques, metaphors, interfaces, etc.) can be remixed with everything'.[5]

Deep remixability is achieved with the new *affordances* of digital media among which *modularity* plays one of the central roles in representational practices. In the context of digital dissertations, modularity undoubtedly deserves more attention as a facilitator of a wide scope of possibilities. Modularity is identified as the 'fractal structure of new media'.[6] It allows for an independence of the parts within digital objects in assemblages. This means that the textual composition, when split into fractal elements and reconstructed as multimodal assemblages, may represent a new segment in the process of codifying systems of transferring an abstract idea into material form.

At the moment, the affordances of digital media permit little more than theorizing about the possibilities of writing with images, sounds and movements in the production of digital dissertations and for general knowledge-construction practices. Thus far, there is no suitable platform that could facilitate user-friendly, widespread engagement of intuitive remixing of text with other representational modalities without prior training in design and the use of specialized creative software. Coming as I do from the field of professional design, the recognition of the sharp dissonance between the possibilities that digital media offer for production of knowledge and their downright limited use in scholarly practice became the main point of concern for my dissertation.

To this end, the rationale for establishing an intuitive digital space where the dynamic embodiment of the contents of a mind with the involvement of signifying multimodal strategies by a scholar or student has first to be justified. In other words, do we really need such a thing as multimodality in writing doctoral or any other scholarly papers? Or is it just the sheer availability of new affordances that encourages trendy tendencies to experiment with the cutting-edge technology?

In the present chapter I would like to give a few examples where I felt I would not have arrived at certain insights without the employment of multimodal semiotics. But I can't do this without briefly discussing

---

5   Vito Campanelli, 'Toward a Remix Culture: An Existential Perspective', in *The Routledge Companion to Remix Studies*, ed. by Eduardo Navas et al. (New York: Taylor & Frances Group, 2015), pp. 68–82 (p. 73).

6   Lev Manovich, *The Language of New Media* (Cambridge, MA: The MIT Press, 2002), p. 30.

Fig. 5 *Fractal structure of new media allows 'deep remixability' of syntagmatic and paradigmatic dimensions in making meaning*, Lena Redman (October, 2020).

three theoretical aspects which, in interplay with digital media, establish the merit of multimodal methodologies in generating, analyzing and presenting research data. These aspects are: *privatized knowledge tools, cinematic writing* and *cinematic bricolage.*

Privatized Knowledge Tools

It is sensible to assume that all contemporary dissertations are digital. To gain access to the global knowledge emporium—that is, to achieve an internationally affirmative evaluation of produced scholarship and thereby acquire an endorsement for its academic circulation—a modern dissertation cannot be anything but digital. As a rule, it should be written in Microsoft Word, presented with PowerPoint, Keynote or Prezi. Its literature review, for the most part, is based on articles retrieved from the internet and eBooks. The discussions and negotiations around the dissertation are often conducted online. This system can be termed *a mainstream digital approach.*

Even though said system makes a solid basis for a modern dissertation to be considered digital, the ontological potential of representational modes and the production of knowledge afforded by digital media remain widely unexplored. As Marshall McLuhan and Quentin Fiore wrote: 'We approach the new with the psychological conditioning and sensory responses to the old'.[7] In the overwhelming 'totality' of digital environments, 'we attach ourselves to the objects, to the flavor of the most recent past' that is, to word-based methodologies, time-approved by traditional academic conventions.[8]

By virtue of inheriting a digital environment—even though many of us are labelled by Marc Pensky's catchphrase as being not 'natives' but 'digital immigrants'[9]—we 'pull ourselves up by our own bootstraps' to become one with the environment.[10] One of the strings in this 'pulling up by the bootstraps' mode of living is striving for the existing media, using McLuhanese, to become 'an extension of man'. Media, as McLuhan and Fiore wrote, 'are so pervasive [...] that they leave no part of us untouched, unaffected, unaltered'.[11]

Having the *privatized knowledge tools* literally in their hands, pockets, bags or in front of their eyes and on their desks, the knower is no longer limited by the knowledge collected and prescribed to them by someone else. The *privatization of knowledge tools* enables the knower to create their own path in the quest for intellectual expansion in accordance with their individual interests, capacities and personal experiences.

---

7   Marshall McLuhan and Quentin Fiore, *The Medium is the Message* (Corte Madera: Gingko Press, 1967), p. 94.

8   Ibid., p. 74.

9   Marc Pensky, 'Digital Natives, Digital Immigrants', *On the Horizon*, 9.5 (2001), 1–6, https://www.marcprensky.com/writing/Prensky%20-%20Digital%20Natives,%20 Digital%20Immigrants%20-%20Part1.pdf

10   Humberto R. Maturana and Francisco J. Varela, *The Tree of Knowledge* (Boston: Shambhala, 1998), pp. 46–47.

11   McLuhan and Fiore, *The Medium is the Message*, p. 26.

Reaching out for the construction of individualized forms of knowledge, the knower, for the first time in history, is afforded an opportunity to reconnect his/her interests with their immediate natural–sociocultural environments, expand and explore this reconnection through a diverse network of cultural resources, and distill the essence of their findings through the expressions of individually and endlessly *hybridized variations.*

Fig. 6 Extension of self-presence in the world by means of digital media, Lena Redman (May, 2016).

Cinematic Writing

The emerging genre, as I saw it, *writing with images, sounds and movements*, has a direct link to the term 'cinematography', as explained by Ed Sikov.[12] *Cinematography* originates from two Greek words 'kinesis (the root of *cinema*), meaning movement, and *grapho*, which means to write or record'. Therefore, Sikov suggested, '*Writing with movement*

---

12   Ed Sikov, *Film Studies: An Introduction* (New York: Columbia University Press, 2010).

and light—[is] a great way to begin to think about the cinematographic content of motion pictures'.[13] In the same vein, *writing with images, sounds and movements* can be termed *cinematic writing*.

In this I find a parallel between the call for an innovative use of the affordances of contemporary digital media—and, in particular, cinematic writing as one of the possible systemic manifestations for knowledge-production—and the invitation to see cinema as an avant-garde means of expression, as suggested by Alexandre Astruc in 1948. 'I would like to call this new age of cinema the age of camera-stylo (camera-pen)', wrote Astruc in his persuasive essay 'The Birth of a New Avant-Garde: La Caméra-Stylo'.[14]

'A Descartes of today', Astruc went on, 'would already have shut himself up in his bedroom with a 16mm camera and some film, and would be writing his philosophy on film [...] of such a kind that only the cinema could express it satisfactorily.' Cinema, according to Astruc, was progressively transforming into a language by which a producer could externalize a meaning 'as he does in the contemporary essay or novel'.[15] Astruc's article was a celebration of the technological potential of cinema to become an expression of an individual mind's most private thoughts—shutting yourself up in your bedroom and responding to the fleeting moments of reality—something that could be conveyed only through the clusters of moving light and shapes entangled in the ever-present intricacy of sounds. Astruc rejoiced in the discovery of a new system of signification, where, as I see it, one mode of expression is just an uttered note, but the cluster of them is a philosophical poem of revealed consciousness.

The process of 'revealing consciousness' by means of merging logic and aesthetics, that is, transitioning from 'mind-cinema' by framing the abstract-implicit into the digital-explicit, became a core point of my study. In that regard, it was important to determine the boundary between the abstract thought and its material manifestation. What exactly is this space through which the transition takes place?

In searching for answers, the analogy of a blind man with a stick, suggested by Gregory Bateson, galvanized my imagination. 'Where

---

13   Ibid., loc. 953.
14   Alexandre Astruc, 'The Birth of a New Avant-Garde: Le Caméra-Stylo', *New Wave Film.com*, http://www.newwavefilm.com/about/camera-stylo-astruc.shtml
15   Ibid.

does the blind man's self begin?' asked Bateson. 'At the tip of the stick? At the handle of the stick? Or at some point halfway up the stick?'[16] The stick, Bateson decided, is a transmitter through which information about the pathway is being continuously broadcast into the man's mind. The man's touching the pathway with the tip of the stick is the primary channel for collecting and transmitting data. The sound that the touch produces, and the sounds, smells and general feelings coming from the surroundings, constitute complementary data-gathering. The senses are engaged in an ongoing collaboration, resulting in a 'systemic circuit' that runs through the stick, maps the pathway in the man's mind, like a movie projector, and reveals new cinema frames associated with each new step that 'determines the blind man's locomotion'.[17]

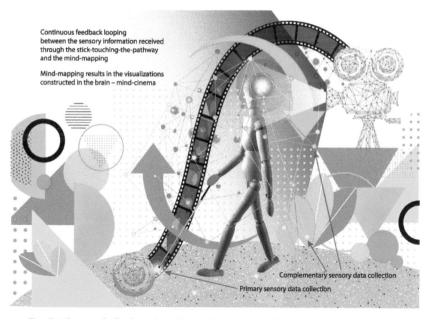

Fig. 7 *A looping feedback produced by the blind man's walking stick touching a pathway and his 'mind-cinema', determining his locomotion,* Lena Redman (April, 2019).

The man, the stick and the pathway, together, are 'a self-corrective unit'.[18] The man's every next step is the result of continuous feedback looping

---

16   Gregory Bateson, *Steps to an Ecology of Mind* (Chicago: University of Chicago Press, 1972), p. 318.
17   Ibid.
18   Ibid., p. 319.

between the information received through the sensory circuit and the mind-cinema playing in his head. This network is not confined by the contours of the head or the body but includes the pathway and the stick through which the information travels, as well as the surroundings, and that is what determines the next move.

Translating this into a meaning-making activity, we can say that every data-organizational and representational move of gathered and analyzed data creates a circuit towards meaning-clarification; and every cleared segment of meaning affects the next step in the organization and representation of data. In other words, the process of ideation is achieved by step-by-step progressive unfolding—a circuit of gathered and analyzed data and self-correcting/self-organizing feedback loops.

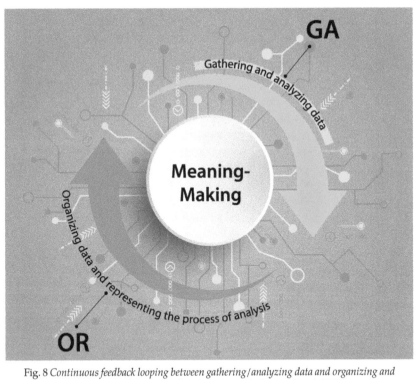

Fig. 8 *Continuous feedback looping between gathering/analyzing data and organizing and representing the process of analysis*, Lena Redman (April, 2019).

This can be reduced to a formula where gathering and analyzing data, positioned in feedback circuits with organizing data and representing the processes of analysis, leads to meaning-making:

Fig. 9 *Symbolic representation of the feedback loops between gathering/analyzing data and its organizing/representing that leads to construction of meaning,* Lena Redman (April, 2019).

So, in what ways has this equation affected my epistemological quest?

Cinematic Bricolage

The equation $GA <> OR >>> MM$—that is, the *Gathering and Analyzing* $<>$ *Organizing and Representing* circuit resulting in the step-by-step production of *Meaning-Making*—gave rise to a question: to what degree does such an approach allow the exploration of the metacognitive function of mind that results in the construction of new knowledge?

The complexity and number of theoretical underpinnings—the structural functionality of semiotics, cinematic writing, the privatization of the means of knowledge production, media as extension of man, and deep remixability—necessitated the imposition of structural clarity, which I found in the adoption of Lévi-Strauss's research methodology that 'is commonly called "bricolage" in French'.[19] The bricoleur, Lévi-Strauss notes, 'has no precise equivalent in English'. She/he is 'a kind of professional do-it-yourself' person who 'uses devious means compared to those of a craftsman'.[20]

The bricoleur mode of operation differs from the engineer's practice. The engineer establishes a question and finds a solution in a planned sequence.

The bricoleur, according to Lévi-Strauss, explores 'the heterogeneous objects of which his treasury is composed to discover what each of them could "signify" and so contribute to the definition of a set which has yet to materialize'.[21] In the context of my dissertation, the 'set that has yet to

---

19   Claude Lévi Strauss, *The Savage Mind*, trans. by George Weidenfield (Chicago: University of Chicago Press, 1962), p. 16.

20   Ibid., p. 17.

21   Ibid., p. 18.

materialize' was an anticipated cluster of insights as to how multimodal semiotics can affect an epistemological progression.

Fig. 10 *Seeing hetero-geneous objects as 'op-erators' in constructing meaning with diverse means*, Lena Redman (December, 2016).

To this end, I decided to explore the representational process of my childhood memories of growing up in Soviet Russia. How do the reorganization of old photographs, objects of personal sentimental value, memories, songs, music, movies from childhood, old video clips from YouTube and other fragments of data organized in a deep remixability technique influence knowledge-related discoveries?

Using multimodality as a 'blind man's walking stick' what would I find that I didn't know before and perhaps wouldn't have found if I didn't use multimodal tools?

Bricolage is an activity that is subject to the given context and a 'universe of instruments [that] is closed'. The game is 'always to make do with "whatever is at hand"'. Accordingly, the bricoleur's toolbox

Fig. 11 *'Discovering' my own past by representing it with the existing objects and memories*, Lena Redman (February, 2016).

and the availability of materials are limited. At the same time, the scope of the tools and materials is heterogeneous in the sense that it is 'the contingent result of all the occasions'. In other words, in examining the use of digital media to write a dissertation, my digital toolbox—mobile phone, iPad, computer and the internet—was limited in the sense that these tools were not specifically designed for my exact area of study, but their utility was based on the principle that 'they may always come in handy'.[22] At the same time, my set of tools enabled a wide scope of representational heterogeneity.

---

22   Ibid.

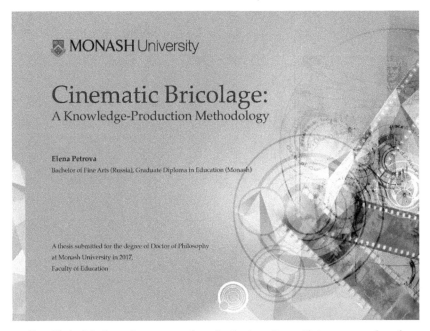

Fig. 12 *A slide-show of some pages from the thesis written with images, sounds and movements in an ePub format*, Lena Redman (February, 2019).

Each page was a composition of multimodal components. Objects appeared and moved in the spaces allocated for them—across the page, behind the text or between the lines; sounds came out, video elements emerged and new text boxes materialized. The focus of analysis was adjusted not so much according to the reception of multimodal compositions by the reader, but rather according to how the engagement with various modes of representation affected the knowledge-producer's cognitive activity.

To avoid communicational complexity, I categorized photographs, scans and other images, music, songs and other database elements, as *bricoles*. Each of the bricoles, borrowing from Lévi-Strauss, 'represent[s] a set of actual and possible relations: they are "operators"'.[23]

What should I expect of the operational functionality of the bricoles in conditions of deep remixability where they are deconstructed and reconstructed in the interplay with other fragments of other bricoles in completely new assemblages?

---

23   Ibid.

Gathering / Analyzing ‹ › Organizing / Representing ›››
Meaning-Making
Looking for Crows

The implementation of multimodal remixes allowed me to notice things that, I believe, would usually be ignored. These were fleeting sensorial moments of a suddenly remembered smell or taste; movements or feelings that my mind produced in response to the awakening of memories. The question emerged: why, when remembering different episodes from my Soviet childhood, did sensorial manifestations—images of crow(s), creeping movements of pipe or cigarette smoke, smells of heavy coats wet from thawing snow—make their sometimes clear, sometimes vague, but nevertheless persistent appearances on the 'screen' of my 'mind-cinema', as if the whole fabric of the screen was woven out of their meaningful relevance?

What if I integrate an animated crow flying behind the typed text and apply the sound of its cawing, mimicking its presence interwoven into my thoughts? How would the conversion of the implicit into the visible and auditory affect the process of meaning-making? Why does it make me 'hear' the Beatles singing:

> Blackbird singing in the dead of night
> Take these broken wings and learn to fly...?

I observed very attentively my mind-cinema responding to my activity working with the multimodal bricoles. I used my mobile phone for taking pictures, and at this stage of the project, crows became my models. I recorded various sounds, and crows' cawing caught my interest. Integrating recorded bits into the pages of writing, I thought that the fact that one of the sound recordings also contained the sound of a random car braking was interesting. I didn't discard it but instead added an image of a car and also incorporated The Beatles' 'Blackbird' song in the background. I tried to avoid elitist thoughts, attempting to block nothing that appeared important even if it seemed nonsensical.

Fig. 13 *Connecting the dots of associations, memories and newly collected information,*
Lena Redman (August, 2016).

A 'whirl of catching and being caught', Tim Ingold wrote in his
philosophically-poetical *The Life of Lines*.[24] In trying to catch meaning, the
meaning 'catches you in your own mind'—that's how I felt in allowing
my thoughts taking digital form.

I observed the dynamic circuit as working on the written text
inspired the generation of certain images, movements and sounds,
and the generation, categorization, curation, and their manipulation
influenced alterations in the written text which, in turn, resulted
in gaining insights. It allowed me to immerse myself in the depth of
memory where some floating fragments were clipped back into their
right places, like disembodied bits of jigsaw puzzles.

---

24   Tim Ingold, *The Life of Lines* (New York: Taylor & Francis Group, 2015), p. 7.

Catching the meaning of the profound sadness of the crow's presence, I understood that it had a deep-rooted cultural significance. The 'black crow' was the term for a vehicle (GAZ M-1) that, during Stalin's time, took people to places of no return. Although the 'black crow' van ceased to exist in the time of my childhood, the term was prominently embedded into cultural representations in movies, books and stories. In my child-mind, the term acquired the literal representation of a bird—a crow.

Smoke Screen

Not less revealing and equally gloomy was the revelation of cultural symbolism in the image of pipe smoke. The sluggishly moving streams spreading behind the layer of typed text were 'catching' something obscure but exceedingly disturbing. 'Being caught' in looking for images of Stalin's 'black crow' van, I became intrigued by the images of Stalin himself and especially by the fact that he was quite often photographed smoking a pipe. There was such an air of significance around the whole thing that it made me even more curious. I started looking for more images and information on the topic.

In Graeme Gill, I read about Victor Deni's illustration that was reproduced in the central Soviet newspaper *Pravda* (1939), where the generals of the White Guard are shown 'being blown away by the smoke of Stalin's pipe'.[25]

I was bewildered by the doctrinal incongruousness of the image. The whole country was heavily conditioned by the Marxist ideology of radical materialism. And here we had the 'Father of the Nation' being portrayed as a magician blowing out some sort of spell on the generals.

But the more images I saw and the more I read about Stalin, the more I came to understand that despite Soviet ideology never tolerating anything supernatural, Stalin himself was portrayed as having superhuman abilities.

It came into focus that Stalin's pipe was actually one of the visual symbols endowing him with 'supernatural power' as he kept brutal control of the country with a population of more than one hundred

---

25   Graeme Gill, *Symbols and Legitimacy in Soviet Politics* (Cambridge, UK: Cambridge University Press, 2011), p. 301.

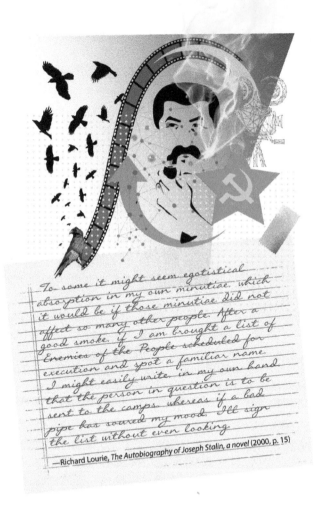

To some it might seem egotistical absorption in my own minutiae, which it would be if those minutiae did not affect so many other people. After a good smoke, if I am brought a list of Enemies of the People scheduled for execution and spot a familiar name, I might easily write, in my own hand that the person in question is to be sent to the camps, whereas if a bad pipe has soured my mood, I'll sign the list without even looking.

—Richard Lourie, *The Autobiography of Joseph Stalin, a novel* (2000, p. 15)

Fig. 14 *Making connections between the crows and pipe smoke*, Lena Redman (May, 2016).

million at the time. In his historical novel about Stalin, Richard Lourie depicts him as an egocentric dictator for whom the quality of a taste of smoke on his tongue would determine the destinies of thousands of people and their families.[26]

The movement in this case, as a mode of representation, was that functional aspect that Lévi-Strauss described as an operator that carried in itself possible relations in recognition of meaning. The pipe smoke mediated Stalin's personality, shrouded in secrecy and ultimate distrust of anyone, and his stealthily well-calculated actions. Not the smoke itself, but its spreading, its continuous thickening, represented heavy oppression and lack of air to breathe for anyone who found themselves under that stealthily moving cloud. Stalin's style of moving and speaking was slow, like his pipe's smoke spreading far away around the whole country, screening the truth and making the cruel lie appear to be a necessity for survival. Like a spider who catches its victims with a hidden web, Stalin dispersed his manipulative smoky nets all over Russia and kept it tight in those nets for years.

Bringing Forth the World Together with Others

The most heartfelt discovery that my multimodal probe provided me with was the epistemological aspect which Humberto Maturana and Francisco Varela described as 'the world that we bring forth with others'. In other words, every act of knowing 'is a structural dance in the choreography of coexistence' with others.[27] To discuss this, I use an ongoing incorporation of—what I felt was the 'appropriate' representation—Beatles songs into the bricolage of probes. While assembling animations to represent my imaginary meeting with 'the fathers of communism', and in particular with Lenin, I felt that juxtaposing the 'Internationale' performed in German and The Beatles' 'Strawberry Fields Forever' was an absolutely right choice in depicting the contrast.

---

26 Richard Lourie, *The Autobiography of Joseph Stalin: A Novel* (Boston, MA: Da Capo Press, 2000).

27 Maturana and Varela, *The Tree of Knowledge*, p. 248.

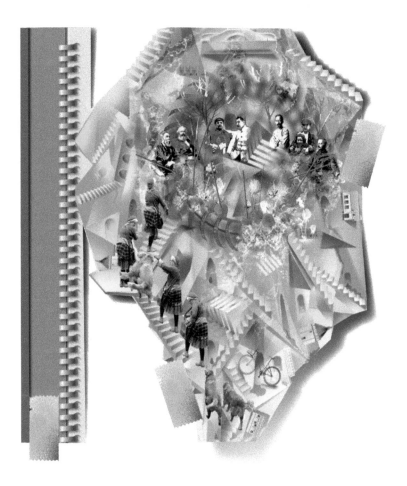

Fig. 15 *Imaginary meeting with the fathers of communism*, Lena Redman (May, 2015).

My intention was to create the sense of a collision by using my childhood's most innocent and happiest moments that, in my particular case, can be metaphorically expressed with the digital 'strokes' portraying a garden, a clear sky, the buzzing of bees and fluttering of butterflies, the fragrance of strawberries ripening in the sun. On the other hand was the harsh austerity and hidden cruelty of Soviet reality. The 'Internationale' sung in German is a reflection of Lenin's 'specific admiration for Germany that was enormous'.[28] Lenin was raised by a mother who was of German and Swedish ancestry and who stayed loyal

---

28   Robert Service, *Lenin: A Biography* (London: Pan Macmillan, 2002), loc. 252.

Professor Woland is a character from the famous Russian book by Michail Bulgakov, 1928-1940, Master and Margarita, a calamitous satire written during the drearisome time of Stalin's repressions and banned until 1967. Woland is an imaginary figure of Satan who visits the Soviet Union.

Fig. 16 *Sanctification of Lenin by the use of fear and oppression*, Lena Redman (April, 2016).

to German cultural traditions.[29] Lenin also lived for a long period of time in Germany and liked the culture. 'He wanted the West too to change. There had to be a European socialist revolution that would sweep away the whole capitalist order'.[30] Lenin appears not to have been very fond

---

29   Ibid., loc. 496.
30   Ibid., loc. 252.

of Russians, who in his opinion, are talented people but 'have a lazy mentality'.[31] The 'Internationale' performed in German sounds to me like the reverberation of Lenin's ultimate goal, in which Russians 'were sorry specimens that populated the corrupt world'[32] and were used as disposable material for the first trial of a larger and more important world scheme.

The images of smashed strawberries, sounds of a battle, and together with them the song 'Strawberry Fields Forever' from the movie *Across the Universe,*[33] which is based on The Beatles' songs, came to my mind as a concrete thread to patch together fragments of simple naiveness and corrupt barbarity. The question that I asked myself was, why did the choice of something to mediate my feelings so often fall on The Beatles' songs?

In the early 1970s my older brother bought on the black market a self-produced *Abbey Road* tape-cassette, paying more than the equivalent of half the average monthly salary. We played it quietly, not to annoy mum and taking care that a censorious ear would not catch the tune.

Earlier still, and under much more dangerous circumstances, I remember the black-market circulation of discarded medical X-ray films with Beatles songs etched on them. Pity I never had one. It could cost you not just two weeks' wages but also your studentship or job, or even worse, depending on the circumstances. A very helpful thing was that those 'bones' records were easy to bend and hide in the sleeve of your coat.

The more information I found on the topic and the more I retrieved facts and events from my own memory, the more I was astounded with the widespread and compelling Beatles subculture in Soviet Russia that I always knew I was part of, but never realized its tremendous cultural significance.

I always knew that my answer to a run-of-the-mill question about what music album I would take with myself to a deserted island would be an immediate *Abbey Road*. The choice, however, is not based purely on aesthetic preferences.

---

31   Ibid., loc. 694.
32   Richard Pipes, *Communism: A History* (New York: Modern Library, 2003), p. 69.
33   Julie Taymor, dir., *Across the Universe* (Sony Pictures Releasing, 2007).

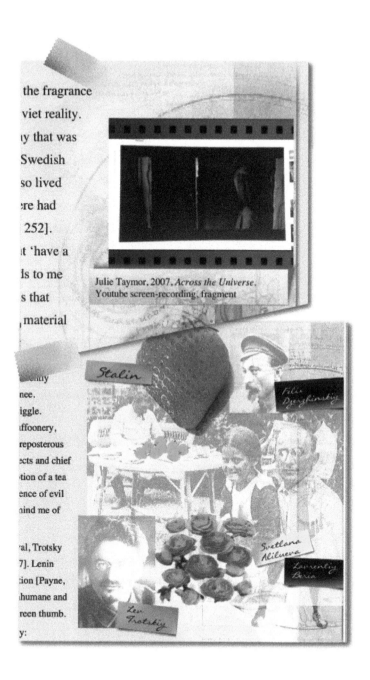

Fig. 17 *'Cut-outs' from the pages of the thesis*, Lena Redman (May, 2016).

Fig. 18 *A record 'on bones'*, Lena Redman (April, 2016).

During my study, not without surprise and excitement, I came across Leslie Woodhead's book, *How The Beatles Rocked the Kremlin*. Bewildered, I read about the parallel universe where my own generation grew up. Many of Woodhead's characters call the Beatles fans (us!) the Soviet Beatles Kids. It was only through my multimodal probing that I came to realize how great a contribution our souls and minds, rocked by The Beatles' music, added to rocking the Kremlin and shaking Soviet Potemkin villages' walls. In the case of the Berlin Wall—literally. We did not even realise that when we (secretly from our parents) were tuning in to the waves of Radio Liberty, catching familiar tunes with poorly recognised English words, and nevertheless *recognised* 'Hey Jude, don't be afraid...', we were joining the vibration of another world, free from smothering ideological coercion. We were getting less and less afraid to stand against the fake façades of Soviet constructions. When our boys were growing their hair like The Beatles, and were very badly treated for this, along with their hair their defiance was growing as well. As one of the Woodhead's characters, Kolya Vasin, recalls:

> 'The policeman said, "You are not Soviet man! You are living like a Western man!" And he grabbed my hair.' The memory of how the cop dragged him along the platform by his hair while dozens of people stared and laughed was branded into him. 'I was almost crying from the pain, but I had to keep silent. I was afraid the man would drag me off to prison'.[34]

Unfortunately, as far as I know, people of that Soviet Beatles Kids subculture never managed to stay back in their home country. Every one among those whom I personally know and would identify as one of the Soviet Beatles Kids, lives at the moment somewhere outside contemporary Russia. As Woodhead observes: 'Millions of kids across the Soviet Union must have shared something of Vasin's despair about their society, strangers in their own country'.[35]

Fig. 19 *The Beatles shook the Kremlin's walls,* Lena Redman (May, 2016).

---

34   Lesley Woodhead, *How The Beatles Rocked the Kremlin: The Untold Story of a Noisy Revolution* (New York: Bloomsbury, 2013), loc. 1886.

35   Ibid., loc. 1052.

It feels as if the Beatles were that voice that made us open our eyes and look critically around. They made us test the ground under our feet and realise that there was no solid substance underneath the artificially constructed surface. It was all just a Potemkin village. This realisation has pushed us out of the country.

Fig. 20 *Inspired by The Beatles, running the Soviet Union*, Lena Redman (September, 2016).

> So we sailed up to the sun
> Till we found a sea of green
> And we lived beneath the waves
> In our yellow submarine...

This is an example of how technology has facilitated the penetration of the Iron Curtain, as Kolya Vasin in Woodhead's book says: 'After the Beatles, the Iron Curtain was like a fence with holes. That was our secret. We breathed through those holes'.[36] The radio played a massive role in this process. Searching for our favourite music on Radio Liberty, we were also given a chance to hear about things in our own country that were hidden from us. These two aspects—the information disclosed about our society and our emotional response to The Beatles' music—were tightly intertwined in our mental schemata. They eventually became inseparable.

The invention of the 'records on bones' was one of the manifestations of growing resistance to the regime. Its symbolical implication, as if we

---

36   Ibid.

were recording 'the spirit of freedom on people's bones', is an indication of the rigor of the subcultural movement. Medical images recorded on film using electromagnetic radiation, together with the technology of etching sounds onto them, made it possible to disseminate the songs.

Conclusion

In those circumstances, with the Beatles' music influencing the young generation of Soviet Russia in the seventies, the technology was a conduit that enabled the flow of independent thought into oppressed reality. Comparing the technological possibilities that were available for the Soviet Beatles Kids with those that are easily accessible for young people today, I think of the great possibilities that education has at its disposal. Maybe the modern system of teaching and learning, designed to guide young people in the 'right' direction and screen them from the troubled world, is in reality the Iron Curtain that prevents them from breathing fresh air, a Potemkin village façade that keeps people from finding out and constructing their own truth about themselves and the world they live in.

In the area of knowledge generation, I see digital media—and cinematic bricolage, as one of its by-products—as an emancipating force available for scholars and students alike to use in their exploration of reality. In addressing this, I join my voice with those of Joe Kincheloe and Shirley Steinberg,[37] Kincheloe (2003)[38] and Kincheloe and Kathleen Berry (2004),[39] who see teachers and students as producers of their own knowledge. Kincheloe and Steinberg argue that self-produced knowledge makes people:

> [...] pursue a reflective relationship to their everyday experiences, they gain the ability to explore the hidden forces that have shaped their lives [...] to awaken themselves from a mainstream dream with unexamined landscape of knowledge and consciousness construction [...][40]

---

37 Joe L. Kincheloe and Shirley R. Steinberg, *Students as Researchers: Creating Classrooms that Matter* (New York: Taylor & Francis e-Library, 1998).

38 Joe L. Kincheloe, *Teachers as Researchers: Qualitative Inquiry as a Path to Empowerment* (New York: Taylor & Francis Group, 2003).

39 Joe L. Kincheloe and Kathleen Berry, *Rigour and Complexity in Educational Research: Conceptualising the Bricolage* (London: Open University Press, 2004).

40 Ibid., p. 3.

Examining beliefs, social practices, dominant standpoints, using materials and tools they have at hand in their given context, and mediating meaning with digital media, cinematic bricoleurs may produce alternative bodies of knowledge. Cinematic bricoleurs may expose the correspondence between the phenomenon they are looking at and the social structures it is embedded in. Throughout this process, they may learn to form their own critical view and find strategies for its advocacy. Speaking metaphorically, they may invent their own 'records on bones' by expressing what previously was obscured from view, using whatever is in their repertoire. This results in the formation of holes in the existing Potemkin villages, making reality easier to see, access, understand and alter when necessary.

The methodology of cinematic bricolage, with its feedback loops of 'catching and being caught', provokes contingent situations and engenders conditions that allow one to take advantage of what was invisible at the start of the project. Such logic was expressed by one of the architects of the Social Study of Information Systems Research, Claudio Cibbora:

> Curiously enough, successful information systems that are developed stem not from formal theories and structured methodologies, or from deliberate designs, but rather from chance events and improvised, serendipitous applications, which are not planned ex ante, and are often introduced by the users themselves through reinvention and *bricolage*; indeed, innovation happens by taking unanticipated paths and timing and assuming a local, apparently inconspicuous character at the outset.[41]

Embracing the process of epistemological innovation with its principle of a self-organizing feedback circuitry, the quotation above can be considered as describing a kinetic force that takes us to the notion of 'churning ripplework', where the circularities push at and pull against each other, causing shape bending and curving, producing multiple overlapping, and thus creating individually hybridized patterns of knowledge.

My train of logic in writing this chapter is tightly intertwined with visualizations and graphic representations and may appear interesting

---

41 As cited in Chrisanthi Avgerou, Giovan F. Lanzara and Leslie P. Willcocks, *Bricolage, Care and Information: Claudio Ciborra's Leagcy in Information Systems Research* (Basingstoke: Palgrave Macmillan, 2009), p. 8.

to some and perhaps incongruous to others, on the grounds that such an approach is incompatible with traditional dissertation writing. Returning to the discussion of the paradigmatic and syntagmatic system of meaning-making, I have to argue that the logic of my approach lies in adding images and graphic elements to a set of traditional signifiers (words), and organizing them in a spatial relation that helps to proceed with very individual way of gaining insights. My argument is not for a professional artistic use of images, sounds and movements—they can be replaced with stick figures, simple shapes, humming and basic moves of the objects within a digital space—but for an analysis of the metacognition that the engagement with multimodality evokes.

My personal 'professionalism' in the use of images and infographics was helpful because it gave me confidence in their employment and, consequently, allowed me to explore the merger of multisystemic signification, where linguistic and visual elements (audio and movements that I used in my dissertation) enabled me to reach, as I believe, the deeper levels of metacognitive processes.

Such knowledge production methodology is oriented towards forging a uniquely personalized repertoire in reaching an intended goal. The cinematic knower/bricoleur starts with retrospection,[42] examining how 'to make do'[43] with the accessible tools and obtainable materials within the given context and according to personal competencies. The knower then works towards the expansion of his/her competencies and skills by connecting fragments of knowledge and their intuitive interpretation, the data from disparate domains, and pulling them into a unique and coherent narrative.

The bricoleur's improvisations are prompted by his/her individual agentic skills in finding resourceful means—or, as Lévi-Strauss puts it, 'devious means'[44]—of negotiating between his/her intention, the availability of resources and skills and the affordances of tools. From this perspective, the methodology of cinematic bricolage emerges not from the professionalism of a craftsman but from a DIY process of organization of fragments from diverse theoretical and practical fields, and by means of the utilization of a uniquely personal range of skills and strategies.

---

42 Lévi-Strauss, *The Savage Mind*, p. 18.
43 Ibid., p. 17.
44 Ibid., p. 16.

Bibliography

Astruc, Alexandre, 'The Birth of a New Avant-Garde: Le Caméra-Stylo', *New Wave Film.com*, http://www.newwavefilm.com/about/camera-stylo-astruc. shtml

Avgerou, Chrisanthi, Giovan F. Lanzara and Leslie P. Willcocks, *Bricolage, Care and Information: Claudio Ciborra's Leagcy in Information Systems Research* (Basingstoke: Palgrave Macmillan, 2009).

Bateson, Gregory, *Steps to an Ecology of Mind* (Chicago: University of Chicago Press, 1972).

Campanelli, Vito, 'Toward a Remix Culture: An Existential Perspective', in *The Routledge Companion to Remix Studies*, ed. by Eduardo Navas et al. (New York: Taylor & Frances Group, 2015), pp. 68–82.

Gill, Graeme, *Symbols and Legitimacy in Soviet Politics* (Cambridge, UK: Cambridge University Press, 2011).

Halliday, Michael A. K., *Halliday's Introduction to Functional Grammar* (New York: Taylor & Francis Group, 2014).

Ingold, Tim, *The Life of Lines* (New York: Taylor & Francis Group, 2015).

Kincheloe, Joe L., *Teachers as Researchers: Qualitative Inquiry as a Path to Empowerment* (New York: Taylor & Francis Group, 2003).

Kincheloe, Joe, L., and Kathleen S. Berry, *Rigor and Complexity in Educational Research: Conceptualising the Bricolage* (London: Open University Press, 2004).

Kincheloe, Joe L., and Shirley R. Steinberg, *Students as Researchers: Creating Classrooms that Matter* (New York: Taylor & Francis e-Library, 1998).

Lévi-Strauss, Claude, *The Savage Mind*, trans. by George Weidenfield (Chicago: University of Chicago Press, 1962).

Lourie, Richard, *The Autobiography of Joseph Stalin: A Novel* (Boston, MA: Da Capo Press, 2000).

Manovich, Lev, *The Language of New Media* (Cambridge, MA: The MIT Press, 2002).

Maturana, Humberto R., and Francisco J. Varela, *The Tree of Knowledge* (Boston: Shambhala, 1998).

McLuhan, Marshall, *Understanding Media: The Extensions of Man* (Corte Madera: Gingko Press, 1964).

McLuhan, Marshall, and Quentin Fiore, *The Medium is the Message* (Corte Madera: Gingko Press, 1967).

Pensky, Marc, 'Digital Natives, Digital Immigrants', *On the Horizon*, 9.5 (2001), 1–6, https://www.marcprensky.com/writing/Prensky%20-%20Digital%20 Natives,%20Digital%20Immigrants%20-%20Part1.pdf

Pipes, Richard, *Communism: A History* (New York: Modern Library, 2003).

Service, Robert, *Lenin: A Biography* (London: Pan Macmillan, 2002).

Sikov, Ed, *Film Studies: An Introduction* (New York: Columbia University Press, 2010).

Taymor, Julie, dir., *Across the Universe* (Sony Pictures Releasing, 2007).

Woodhead, Lesley, *How The Beatles Rocked the Kremlin: The Untold Story of a Noisy Revolution* (New York: Bloomsbury, 2013).

Young, Cathy, 'Stalin's Allied Atrocities' (March 12, 1990), *Washington Post*, https://www.washingtonpost.com/archive/lifestyle/1990/03/12/stalins-allied-atrocities/3855ede8-597f-45fd-84b3-e32d1dba59ad/

14. Precarity and Promise

Negotiating Research Ethics and Copyright in a History Dissertation

Celeste Tường Vy Sharpe

Digital history has long been associated with George Mason University (GMU). The pioneering work of Roy Rosenzweig and many others in establishing the Center for History and New Media in 1994 (now the Roy Rosenzweig Center for History and New Media, hereafter RRCHNM) has put GMU on the map as a leader in digital work. Surrounded by the innovative digital work at RRCHNM, and encouraged by the required two course sequence of digital theory and praxis in the history PhD program, I conceptualized my dissertation project as a web-based work of scholarship. As a history of twentieth-century American visual culture and disability, I envisioned my dissertation benefiting from the multimodal potential of a digital presentation for visually providing layered analyses and interpretations of visual artifacts. That potential is tempered by precarity, by a complex set of factors, power and relationships for a doctoral student to negotiate.

Centering visual materials as objects of analysis in digital work raises unique questions around access, presentation and scope of work at multiple levels. Effectively pushing beyond the convention of images as decorative illustrations within a dissertation requires education and resource support for the doctoral student, the degree-granting institution, university presses and the discipline. Without concerted effort at each level to better deal with issues of copyright and image permissions, doctoral students pursuing visual culture oriented digital

https://doi.org/10.11647/OBP.0239.14

dissertations will continue to be overburdened with complications for their scope of inquiry and career prospects. This chapter will first introduce my dissertation project, then discuss the particular issues I encountered during the dissertation as illustrative examples of questions facing scholars whose source bases feature visual materials (photographs, posters, film and television clips) that may fall under copyright. It was these conversations and negotiations around image and media permissions that created instances of precarity and opportunity with the organizational archives I consulted, with my university, and with my discipline.

The dissertation, 'They Need You! Disability, Visual Culture, and the Poster Child, 1945–1980', examines the visual history of the national poster child—an official representative for both a disease and an organization—in post-World War II America. I argue that the poster child fundraising and educational campaigns deployed by the March of Dimes (MOD) Foundation and Muscular Dystrophy Association (MDA) shifted American understanding of physical disability in new ways by capitalizing on mass media to depict disabled children within their families and communities as full, if physically limited, citizens of the nation. These campaigns used the rhetoric of disease eradication to promote medical cures for disability and illness, and highlighted their role in funding these cures. The organizations' approach promoted a narrow view of disease and disability as largely childhood conditions to be overcome, which also precluded political avenues and policies beyond medical research such as family support and advocacy for disabled adults experiencing post-polio syndrome or continuing to live with muscular dystrophy.

'They Need You' is presented through the publishing platform Scalar, created and maintained at the University of Southern California. The project is organized and presented through a visual index which appears on the project home page. The goal of the visual index is to make visible each project element: image, textual source, analytical writing, annotation, historiography, etc. This allows readers to explore content through a set of themes or keywords (such as gender, family, March of Dimes); hovering over one element highlights its connections to themes and to other elements. I created six 'paths', Scalar terminology for a series of linked elements, that create linear reading experiences through the content.

My rationale for pursuing a fully-digital dissertation is grounded in the fields framing this project, particularly visual culture studies and disability studies. I wanted to emphasize the centrality of the visual sources, and to make those analyses apparent to readers. While print conventions allow for images to be interspersed in text, I wanted to push beyond static placement and the inability to present multiple interpretations or narratives. Additionally, a layered visual presentation would begin to address the reality that disabled children have been historically silent, and silenced, in the archival record. Children's voices have not been well-documented, and disabled children have historically been institutionalized, marginalized or erased by the adults surrounding them.[1] Surfacing these lived experiences and voices in the archive is vital. I sought to avoid re-inscribing the silences by highlighting disabled children in these organization's collections and using the affordances of digital publishing platforms to make these artifacts available for other researchers and interested readers.

The tenuous state of both archives made it imperative to me to make visible as much of the source base as possible to ensure the collections could continue to be used in some capacity. Over the last two decades, the MOD and the MDA had gradually reduced funding and staffing for their archival collections. Neither organization had fully catalogued and processed their collections, and neither had they done any digitization beyond a handful of images requested and paid for by documentarians and previous researchers. The MOD's reduction occurred more recently. Until 2015, the organization's sense of historical memory justified dedicating resources to the preservation of the archive and opening it in some capacity to researchers. Two key events shaped the charity's desire to keep and maintain an archive: the organization's founding by President Franklin Delano Roosevelt, and the successes of their early work on polio and namely support of the Salk vaccine. As a result, they prioritized processing the collections most relevant to FDR and the medical story of developing the cure for polio over the collections from the education and women's division among others. The MDA, on the other hand, laid off their archivist years ago and moved much of their

---

1   Susan Burch and Ian Sutherland, 'Who's Not Yet Here? American Disability History', *Radical History Review*, 94 (2006), 127–47 (pp. 128–30), https://doi.org/10.1215/01636545-2006-94-127

historical materials into a non-temperature-controlled storage unit near their headquarters in Tucson, Arizona.[2] The materials available from the MDA were primarily published reports and newsletters, and I was given access to approximately five unprocessed boxes of historical materials. These boxes turned out to be full of irreparably heat damaged artifacts: photographs melted together, and film strips loose and sandwiched between stacks of papers. The MOD and MDA are not alone in putting scant resources toward the preservation of their historical materials, and many of the challenges digitizing those collections are well-described by archivists and librarians in public institutions.

Private organizations have their own senses of institutional memory, and are not immediately inclined toward supporting access to their materials. Unlike working with academic libraries and archives— or private research institutions such as the Huntington Library or Rockefeller Archive Center—active charitable organizations are focused more with their current mission and public perception and reputation, and can take wildly different approaches to their holdings. Both the March of Dimes and Muscular Dystrophy Association take a commercial and protective approach to their archival materials. For the MOD, popular interest in FDR and the story of polio in America has created an opportunity for the organization to profit from the licensing of their archival materials. As a result, I found them more interested in asserting their claims to copyright than in discussions around fair use or the academic use of these materials.

The MDA has a different attitude toward their archival materials. The vast popularity of Jerry Lewis in their fundraising activity, and the subsequent backlash against his attitudes and performance by members of the disability rights movement, has led the MDA to take a more cautious stance with regards to researchers and external interest.[3] Whereas the MOD recognizes, to some extent, their place in the nation's historical memory, the MDA is more narrowly focused on current public perception. For example, a MDA staff member expressed concern over my interest in a 1950s era photograph depicting a firefighter teaching

---

2   Information told to me in conversation with two MDA officers. I was unable to get access to internal documents or published records to verify the information.

3   Paul K. Longmore, 'The Cultural Framing of Disability: Telethons as a Case Study', *PMLA*, 120 (2005), 502–08 (p. 505), https://doi.org/10.1632/s0030812900167793

children using wheelchairs about fire safety. Their point of concern was the firefighter's use of graphic photos of burned bodies as evidence of the dangers of fires for wheelchair users. While they allowed that I understood the context around the image, they worried about an uncritical viewer's negative response toward the firefighter. And the firefighters' union, I learned soon after, was one of the first and most enduring group of supporters for the charity. The MDA staff member tried to redirect my interest toward a less contentious image, and further, cast doubt on whether I could include the original image or if they could assert copyright to put the photo out of my reach. The MDA's reaction to the firefighter photo highlights their wish to more directly curate and shape not only access to the materials, but also the resulting analysis and presentation that would result from my research.

The charities' attitudes raise serious questions for researchers in general, but particularly for doctoral students who likely lack the institutional support and resources of full-time academic faculty and staff. My status as a doctoral student, and the digital nature of my dissertation, complicated the organizations' responses to my inquiries. Overall, they understood that I was an individual researcher-in-training, with limited funds, and were generally sympathetic to what I was trying to achieve. The March of Dimes archivist, and the art director and VP of Publications for the Muscular Dystrophy Association, all responded kindly to my requests for access and were helpful in locating relevant material. However, those individual kindnesses were tempered by an organizational wariness toward anyone who could potentially cause tension or complicate the narratives they wished to tell about their own organizational histories. The digital aspect of the dissertation was at once the most interesting and the most fraught aspect of our interactions. Here, tensions arose between individual staff member and the larger organizational views of what it means to have digital artifacts and information online for external audiences.

Copyright is a well-known landmine when pursuing digital work. In the field of United States digital history, the majority of projects either pursue pre-copyright topics (pre-1923) or center on digitized public domain sources from the twentieth century. Projects that have managed to partner with or gain access to private collections, such as the *Robots Reading Vogue* project, rely on institutional reputation and resources in negotiating with private companies for access to historical materials.[4] A lone doctoral student, on the other hand, lacks those supports and is in a more vulnerable position in dialogue and negotiations with private organizations.

Like many universities, GMU places the onus for copyright clearance and use rights on the student. Institutional training and support in these areas, however, are thin. At the department level, the formal digital dissertation guidelines mentions issues in copyright and use agreements in the context of the bibliography and the need for a statement describing the factors behind whether the project data is publicly available or not.[5] Copyright and fair use were taught in the abstract during the required digital coursework, and was only rarely an issue for RRCHNM projects, so the department had never had to grapple with their role and responsibility for training students to negotiate these issues. The library was also little help. The copyright expert in the scholarly communications department of the library pointed me toward some external resources, and was able to provide only general advice and otherwise referred me to legal counsel. The latter suggestion was never a viable option for me financially, and I was unsure how to follow up on my questions or how to seek external advice. Neither the history department nor GMU are unique in this regard: the complexity and the legal aspects of copyright and fair use are fraught and often seen as beyond the scope of a department/library/college's purview. While understandable to some degree, the result is that doctoral students are ill-prepared to consider the ethical and legal questions of intellectual property that are increasingly prevalent in pursuit of scholarship in the digital world. Left to themselves, success is largely driven by an individual doctoral student's ability to seek and to avail themselves of external resources and contacts. Shifting some of that responsibility and support to the degree granting institution, from the department to the library to the graduate programs, would help alleviate the burden on the student, who is unquestionably the most vulnerable and precarious actor in this process.

---

4   Lindsay King and Peter Leonard, *Robots Reading Vogue*, http://dh.library.yale.edu/projects/vogue/

5   Department of History and Art at George Mason University, 'Digital Dissertation Guidelines', https://historyarthistory.gmu.edu/graduate/phd-history/digital-dissertation-guidelines

The ubiquity of information online and the ways in which content is constantly sampled, remixed and decontextualized forced the organizations to consider their own digital content strategies and the extent to which they accepted scholarly work entering the equation. A public-facing digital dissertation, like it does for academia, raised uncomfortable questions for both organizations. The limited circulation of a print dissertation was a familiar convention, and allowed the charities to take a generally more tolerant attitude toward the reproduction and use of their archival materials in that context. But a project accessible to anyone with an internet connection, one that could be viewed by a wide range of people, was an entirely different proposition. The presentation of the visual artifacts themselves was the key pain-point; the charities would have preferred the convention of describing images through prose, with only a handful of carefully selected images visible. In their view, the images-as-illustrations model mediates some of the perceived risks of making materials available online by allowing the charities (and perhaps scholars) to curate and present only those images that pass muster across a range of criteria: availability, content, copyright status.

The perceived risks associated with scholarly publication of archival materials can be seen most directly in the fees assessed. In a 2015 conversation with the MOD, I was given the rate card with all charges listed: the cost per image for a scholarly publication (book or article) was $50; the digital rights to the same image cost $500.[6] At the time of this conversation, I had identified 120 images that I definitely wanted to include in the project and potentially hundreds more. A base cost of $60,000 is an unbelievable amount for any work of scholarship, let alone a dissertation. The MDA did not share their fee schedule with me, but did intimate that these decisions were made in consultation with their legal team. In fact, staff at both organizations referenced the presence and tenacity of their legal team in protecting the organizations from malfeasance. In the end, I avoided paying for image use by agreeing to password protect the images to avoid the wider dissemination that the charities worried about. I was disappointed with this outcome, but without significant personal resources or greater institutional support I felt that I was unable to negotiate further, and was fearful that any misstep would embroil me in a legal dispute that I could not win.

---

6   March of Dimes Foundation, Rate Card, 2015.

This experience is illustrative of the risks doctoral students can face pursuing a public-facing digital dissertation, and the complex trade-offs necessary to complete a project so reliant on visual materials. It is common advice to avoid such knotty topics in the first place, to choose a project where the challenges are well-known by faculty advisors and relatively benign. And to some extent, such advice is sound. The larger implication, however, is that doctoral students—and by extension their faculty mentors—do not get the opportunity to learn and expand their knowledge on these questions by working through such difficult problems programmatically. At a higher level, historical inquiry as whole is impoverished by the repeated attempts to neatly avoid the messiness of copyright, fair use, intellectual property, and privately held organizational archives.

The charities and I had to contend with a challenging question: to what extent was a digital dissertation available online a publication? It could be argued that a print dissertation in the North American context is a quasi-publication, with a limited intended dissemination and, in disciplines like history, serves as a dress rehearsal for a monograph.[7] As such, it is constrained in its circulation by design. But to the charities, a dissertation presented online was a publication by virtue of being online. Whereas the nuance between self-publication and pee- reviewed publication, and the emphasis on the latter as conducted by a recognized press as a legitimizing feature, are known and fairly well understood in academic circles, those distinctions meant little to the organizations. While a publicly-available digital dissertation seemed to be a publication that merited the same fees as any other website or documentary request to the charities, its meaning within the pipeline of academia is less clear-cut.

Publication pipelines for digital projects and monographs are still emergent, and continue to lean heavily on the print paradigm. One example is found within the primary style guide of the discipline, *The Chicago Manual of Style*. The guide perpetuates some of the conceptual difficulties through its continued categorization of visual sources as 'illustrations', and assignment of a separate treatment from the

---

7   Fredrika J. Teute, 'Dissertations Are Not Books' (April 1, 2015), *Perspectives in History*, https://www.historians.org/publications-and-directories/perspectives-on-history/april-2015/dissertations-are-not-books

note and citation format of textual sources.[8] This distinction and the subsequent conventions around captioning and inserting 'illustrations' into text implicitly renders a value judgment that images are separate at best, and decorative at worst. That attitude toward images has strong influence on how disciplinary departments, university libraries and presses, and even private archives view the study and presentation of visual materials. Throughout the dissertation stage, I heard variations of the same question from people in all the above listed groups: could I substitute public domain images for the ones held in private collections? This notion of changing images makes sense for those for whom images are illustrative but not integral, and in select cases where another image might be able to display similar content or composition to help support the argument. In a project where the specific images are the central objects of study, substitution substantively shifts the inquiry by changing the core source base. Thus, the implications for increasing institutional support at multiple levels for public-facing, image-centered digital work touch the direction of scholarship for the discipline as a whole.

Copyright and image clearance in a digital ecosystem is a thorny issue, one publishers have so far hesitated to take on. If the March of Dimes is a warning indicator, fees for digital rights now significantly exceed the typical subvention costs associated with publishing images. The institutional weight and reputation of the press can factor both positively and negatively: for every bit of legitimacy lent to the project and author, there is also the institutional risk and cost calculus to take into account. Responsibility for those costs and agreements still largely falls to the author, though an author under contract occupies a more secure position from which to negotiate than a doctoral student. The implications for this tension resonate back to the choice and definition of a dissertation project's scope of inquiry. In a monograph-oriented discipline like history, the prospect of pushing a project through from dissertation to published monograph is a significant factor. Without explicit support from publishers, an image-heavy project has a limited and precarious lifecycle during and beyond the dissertation stage, which can have serious ramifications for emerging scholars who wish to study visual artifacts and analysis in digital spaces.

---

8   The University of Chicago, *The Chicago Manual of Style*, 17th edition (2017), http://www.chicagomanualofstyle.org/book/ed17/part1/ch03/toc.html

Without increased and concerted institutional support from department to university to publisher, digital dissertations will continue to be a potentially precarious choice for doctoral students. Broader engagement with issues around copyright, fair use and intellectual property at all levels—departments, universities, disciplines and publishers—can more fully open opportunities for digital dissertations to take greater advantage of the affordances of the medium and continue to widen the avenues and kinds of research pursued.

Bibliography

Alliance for Networking Visual Culture, *Scalar*, https://scalar.me/anvc/

Burch, Susan, and Ian Sutherland, 'Who's Not Yet Here? American Disability History', *Radical History Review*, 94 (2006), 127–47, https://doi.org/10.1215/01636545-2006-94-127

Department of History and Art at George Mason University, 'Digital Dissertation Guidelines', https://historyarthistory.gmu.edu/graduate/phd-history/digital-dissertation-guidelines

King, Lindsay and Peter Leonard, *Robots Reading Vogue*, http://dh.library.yale.edu/projects/vogue/

Longmore, Paul K., 'The Cultural Framing of Disability: Telethons as a Case Study', *PMLA*, 120 (2005), 502–08, https://doi.org/10.1632/s0030812900167793

March of Dimes Foundation, Rate Card, 2015.

Teute, Fredrika J., 'Dissertations Are Not Books' (April 1, 2015), *Perspectives in History*, https://www.historians.org/publications-and-directories/perspectives-on-history/april-2015/dissertations-are-not-books

The University of Chicago. *The Chicago Manual of Style*, 17th edition (2017), http://www.chicagomanualofstyle.org/book/ed17/part1/ch03/toc.html

15. Lessons from the Sandbox

Linking Readership, Representation and Reflection in Tactile Paths

Christopher Williams

In December 2016, I successfully defended a natively digital dissertation in Artistic Research entitled 'Tactile Paths: On and through Notation for Improvisers'[1] (hereafter, TP) at the Academy of Creative and Performing Arts, Leiden University. In this chapter I will share a few field notes on my motivation for publishing TP as a website, my chosen platform, on the collaboration with a professional designer, and the consequences of this route for the research process itself.

I write as a user of digital publication tools, rather than a scholar or designer of them. Both despite and by virtue of this limited perspective, I hope what follows might prove a useful case study for writers and theorists of digital dissertations, in Artistic Research and beyond.

Topic

The topic of TP is the encounter of notation and improvisation in contemporary music, of which I am both a scholar and a composer-performer. This topic was a crucial factor in my embrace of the web-based format. Historically speaking, notation for improvisers is nothing new. There are many examples of music across epochs and cultures in which notation and improvisation fruitfully coexist: Tibetan ritual

---

1   Christopher Williams, 'Tactile Paths: On and through Notation for Improvisers' (PhD dissertation, Leiden University, 2016), http://www.tactilepaths.net/

    https://doi.org/10.11647/OBP.0239.15

horn tablature, Duke Ellington's big band music, or the Chinese guqin tradition. Up until the late nineteenth century, this encounter was also common in Western concert music, e.g., medieval chant, baroque *basso continuo* and virtuosic Romantic solo repertoire. However, throughout the late nineteenth and early twentieth centuries, academic discourse and compositional practice in Western concert music tended toward the idea that written notation should comprehensively prescribe and/or preserve the salient elements of musical works. Consequently, improvisation became less visible than it had been previously.

Over the last fifty years, improvisation has returned to prominence in both practice and theory. Many key figures in contemporary music have integrated both notated and improvised practices into their work. Likewise, performativity and improvisation have begun to recapture scholars' attention. Nevertheless, most contemporary music that works explicitly and dynamically with improvisation and notation still remains underrepresented. Very little of this work has been analyzed collectively. The goal of TP is thus to articulate a conceptual framework for these practices. To this end, I analyze aesthetically and historically diverse examples of the practice to find common methodological threads. These threads reveal core issues and principles applicable to a broad spectrum of other work, trends and problems in contemporary music.

Field

TP inhabits the broader academic field of Artistic Research (hereafter, AR). Like the topic of TP, a few words on AR should help clarify my approach to native digital publication. Henk Borgdorff, an oft-cited AR spokesman, characterizes the field as follows:

> The expression artistic research connects two domains: art and academia. Obviously the term can also be used in a general sense. Every artist does research as she works, as she tries to find the right material, the right subject, as she looks for information and techniques to use in her studio or atelier, or when she encounters something, changes something or begins anew in the course of her work. Artistic research in the emphatic sense [...] unites the artistic and the academic in an enterprise that impacts on both domains. Art thereby transcends its former limits, aiming through the research to contribute to thinking and understanding; academia, for

its part, opens up its boundaries to forms of thinking and understanding that are interwoven with artistic practices.[2]

[A]rtistic research—embedded in artistic and academic contexts—is the articulation of the unreflective, non-conceptual content enclosed in aesthetic experiences, enacted in creative practices and embodied in artistic products.[3]

Following Borgdorff, I seek to articulate 'the unreflective, non-conceptual content enclosed in aesthetic experiences'—to unpack my own practical work with notation for improvisers. Since existing theories of notation and improvisation fail to explain the research objects, for reasons described above, working with the materials firsthand is a useful point of departure. Strategic documentation of this practice provides rich data that more conventional scientific observation may be ill-equipped to capture. At the same time, reflection in the process of documentation deepens and transforms the practice: observing oneself working changes how one works. This is not merely a condition of Tactile Paths; it is an objective. In the dissertation I aimed to develop my own creative work and that of my artistic community, and thus 'unite the artistic and the academic in an enterprise that impacts on both domains' (my emphasis). In other words, Tactile Paths is both a discussion and an instance of artistic practice.

Motivation: Readership

Early in my doctoral studies, my supervisors Marcel Cobussen, Richard Barrett and Frans de Ruiter suggested publishing TP as a website. Cobussen, author of one of the first natively digital PhD dissertations[4] in Europe, was especially encouraging.[5] His rationale was twofold. First, readership would increase immensely. As an example, he cited regularly

---

2   Henk Borgdorff, 'The Production of Knowledge in Artistic Research', in *The Routledge Companion to Research in the Arts*, ed. by Michael Biggs and Hendrik Karlsson (New York: Routledge, 2011), pp. 44–63 (p. 44).

3   Ibid., p.47.

4   Marcel Cobussen, 'Deconstruction in Music' (PhD dissertation, Erasmus University, 2002), http://www.deconstruction-in-music.com/navbar/index.html

5   I should note that Cobussen himself received no such encouragement as a student. Like editor Virginia Kuhn, he encountered stiff resistance from university administration. For having 'broken the digital ice', I am grateful to him, Kuhn and others featured in this volume.

receiving emails from new readers of his dissertation nearly fifteen years after its publication. On this point I was immediately sold. In terms of accessibility, a bound volume housed in a university library or a PDF stored in academic databases can hardly compete with a contemporary open-access website. Like scholars in any field, I wished to maximize exposure. Specifically, I wished to reach fellow practitioners who would rarely, if ever, find my work in academic contexts. Many of my peers in contemporary music read professional magazines and books, but only a limited number of them read academic journals where TP might be cited or published in article form. Even fewer would find TP directly in a university library or digital repository. A website would intersect with this demographic's habitual knowledge circuits.

Cobussen's second rationale was the ease with which the web absorbs nonverbal media such as audio, video, and musical scores. Integrating such materials is far less practical with physical media. Examining a large, detailed score in a bound A4-sized book while navigating a separate DVD can be clumsy. Comparing multiple audio recordings simultaneously is impossible from a single playback device. On a website, however, such functionalities are painless for the writer to implement and the reader to engage with. This ease of media integration was important for presenting work by other artists that I would analyze. Since the practices under scrutiny were poorly represented in mainstream publishing and scholarly discourse, readers needed access to high-quality primary sources. I also needed to adequately represent my own creative work at the heart of the research. In both cases, showcasing time-based and nonverbal media was essential.

Discovering Design Issues: Representation

After deciding to publish TP in a natively digital form, a website, my next step was to investigate the *Research Catalogue* (hereafter, *RC*), a go-to platform for many artistic researchers. *RC* combines a website-building interface, hosting service, and social network for diverse AR projects and publications. Users include over 10,000 graduate students and professors, independent scholars, institutions and journals such as the *Journal for Artistic Research*, *RUUKKU* and the *Journal of Sonic Studies*. Initially, *RC* seemed ideal. Cost-free membership provides simple, intuitive tools for building multimedia expositions; unlimited media hosting and high-quality media players; basic technical support from *RC* staff; and access to a large AR community. After further experimentation, though, I found a number of fundamental problems. First, *RC* layouts are non-responsive. They do not adapt to differently sized computer screens, tablets, and smartphones.[6] On smaller devices, all visual information shrinks to fit the screen. This is an obvious problem for fatigued eyes. While one rarely reads an entire dissertation from a smartphone, a casually interested reader may initially peruse it this way. If the contents are illegibly small, one may not return to it later on a larger device.

Second, *RC* links styling and content in raw HTML; it does not permit CSS.[7] Authors must manually place and format every section of a text and multimedia object in a given project. This work then expands by a factor of (x) number of pages in a given project. (One can imagine the stress this creates for designing a dissertation of several chapters.)[8] *RC* thus provides no centralized way to ensure consistency among layouts and typography classes across an entire project. Furthermore, sharing and forking layouts is impossible; this is surprising, considering the community-oriented nature of the initiative.

The third problem I discovered is a consequence of the first two: all content in RC, including thumbnails and media players, occupies an absolute position in the layout. Sticky headers and sidebars, or areas of a page that remain in place at the top or on the side of the screen independent of user scrolling, are impossible to implement. If an

---

6   *RC* founder Michael Schwab states that '*RC* was conceived and developed by SAR [Society for Artistic Research] at a time when the internet was still very much a desktop affair [2010–12]. This explains the above mentioned developments [non-responsive layouts], but also the slowness in which things happen: SAR is not a business with investment and return, but a non-profit association that serves its members (and the wider community). So I won't be able to say much about how things will look in the near future regarding issues of responsive design etc'. Personal email to the author, June 23, 2018.

7   Cascading Style Sheet is a standard styling language used by web designers to format HTML content. It permits global changes to layouts, typography, etc. To their credit, the technicians at *RC* did attempt to test the integration of a CSS template at my behest. Unfortunately, it was not successful.

8   Schwab notes that 'the publishing of theses may have been in people's minds, but did not matter at this stage. What mattered was to create a link between sustainability, a requirement for serious publishing, and (web) design'. Schwab, personal email to the author, June 23, 2018.

exposition contains a lengthy scrollable verbal text, media players and thumbnails usually occupy a fixed location within or beside it. Visually, this relegates nonverbal media to the status of figures or footnotes, i.e., supporting evidence. On the other hand, nonverbal media or players can be large and centrally placed, with verbal text placed in an internally scrollable window beside them. In this case, the hierarchy is reversed, and verbal text may resemble a caption or program note. This problem turned out to be a dealbreaker for TP. In order to present nonverbal media on equal footing with verbal text, I wished to make all materials as accessible as possible to the user/reader at all times. This would prioritize primary sources as I required and allow for quick and easy cross-referencing.

Lesson #1: Nuts and Bolts Help Reveal the Big Picture

While it may sound as if I came to *RC* expecting things it could not offer, the truth is quite the opposite. Before experimenting with *RC*, my understanding of responsive design and CSS was hazy at best; I did not even know sticky elements existed. I came to see the need for them through my frustration with *RC*'s structural inflexibility. Ironically, working through this technical frustration gave a boost to my research process as a whole. It was clear before addressing any design questions that verbal texts and nonverbal media should intertwine in a general sense. However, the particulars of their integration only began to emerge through experimentation. My principal discovery, as elementary as it was consequential, was that multiple media files could be grouped as a gallery independent from the verbal text. The possibility of emphasizing scores, video, and audio together in the layout highlighted the exhibitional potential of the web format in a way that I had not imagined. That potential, combined with my original intention to share my research objects with a broader public, gave the dissertation a new curatorial impulse.

Beyond considering how to package my arguments effectively for my supervisors and a few selected experts, I began to grapple with how to represent my field of inquiry appropriately. Choosing my objects of analysis was no longer merely a function of theoretical relevance or personal taste, but also of their potential to provoke readers to rethink the identity of my topic. I believe the final constellation of research objects, i.e., the work I discuss in Paths 1–5 (a Path is the equivalent of a chapter in this website), reflects this new concern. The music analyzed is radically distinct on aesthetic, historical and methodological levels. The scores employ a range of notational elements: from informal verbal instructions to detailed conventional Western notation and abstract graphics. Some of the work is more exemplary of notation for improvisers, some less obviously so.

The website also became a space for users to explore materials themselves. To this end, interaction between verbal text and multimedia from Path to Path is also quite diverse. For example, in Path 1, an original documentary film plays a leading role in articulating physicality in works by Malcolm Goldstein; the brief verbal text serves simply to frame it. In Path 3, the verbal text is the protagonist, but a significant portion of it mixes analytical reflections with playful performance instructions reminiscent of the work under lens, Ben Patterson's *Variations for Double-Bass* (1961). Path 5 presents a collection of unpublished scores and out-of-print recordings by Bob Ostertag in a grid that allows readers to trace how notation and recordings in a four-part project relate over time.

Granted, some of these results may have been possible in book form, and/or for an exclusively academic audience. My point is that the very need to think about these issues of representation took root in a technical discovery, sticky media galleries and the disseminatory affordances of an open-access website more generally. Without tinkering with the nuts and bolts of web design, this would not have happened.

Implementation: Reflection

Presented with the above-mentioned arguments for using an alternative to *RC*, my supervisors agreed to fund professional web design. I chose to work with Patricia Reed of Leaky Studio, a specialist in projects for art, culture and academia. She played a key role in maximizing TP's potential, both content- and audience-wise. Reed proposed developing TP with WordPress (hereafter, WP), a free[9] and open-source content management system used to build an estimated 25–30% of the world's

---

9   Unlike RC, WP does not offer free hosting.

websites.[10] Aside from Reed's own positive experience with the platform, two specific features recommended WP for the project. First, WP uses CSS, along with Java and other common web scripting languages. This permitted responsive layouts, sticky sidebars and other basic functionalities I missed from *RC*. WP and CSS decouple content from styling, so non-expert users such as myself can input their own content when working in tandem with a programmer/designer who customizes front-end and back-end to their needs. Second, there is a community of millions of developing themes, improvements, plugins, add-ons and security fixes for WP mainly free of charge to implement. (This is a stark contrast to *RC*, which limits authors to its own tools.) As WP becomes more prolific, the likelihood of rapid software irrelevance is improbable. Although WP was initially conceived as a blogging platform rather than a durable archive, its ubiquity has ironically translated into a kind of de facto 'security or permanence factor', according to Reed.[11]

Using WP for a dissertation has its limitations, as does any platform. First, decoupling content from styling is a double-edged sword. Authors who are unable to do their own specialized design work—more the rule than the exception in a field such as music—need support for all but the crudest tasks. Generally this costs money. I had the rare privilege of support from a forward-thinking department that was both willing and able to pay Reed for her labor. Most PhD students, particularly in our neoliberal era, do not. Technical interdependence in WP also has its downsides. In order to solve a variety of aesthetic and functional issues, we used plugins by third-party developers. Compatibility among these plugins and with rolling WP system updates requires maintenance. Reed states that 'all existing uploaded/created content must remain updated within the system—meaning that yearly upgrades would be suggested, as opposed to the printed dissertation which can be done once and forgotten. Online is synonymous with perpetual attention (upgrades)'.[12] This is precisely the sustainability problem that *RC* solves by limiting design tools to its own.

---

10 Monty Munford, 'How WordPress Ate the Internet in 2016… And the World in 2017' (December 22, 2016), *Forbes*, https://www.forbes.com/sites/montymunford/2016/12/22/how-wordpress-ate-the-internet-in-2016-and-the-world-in-2017/

11 Personal email to the author, June 23, 2017.

12 Personal email to the author, June 23, 2017. At the time of writing, eighteen months after defending TP, I have not updated any element of the website. It looks the same as when it was finalized.

Collaboration

The interpersonal element was at least as important to TP's design as conceptual and technical factors. Reed and I began working together in early 2015, about two years before I defended TP. The time we spent collaborating while I was writing the content allowed her knowledge of design to enrich my own research process.[13] Our point of departure was the following outline, which I prepared alone in October 2014:

Tactile Paths — WEBSITE STRUCTURE

Introduction/ Table of Contents + TAGS page

- "Chapters / TAGS" button organizes info at top of page like http://charliemorrowevents.blogspot.de/?view=flipcard

- TAGS are topics that apply to multiple chapters

- clicking on each TAG opens a window with explanation

- clicking on chapter title goes to that chapter

12–15 "chapter" pages (each about a single musician/ group/ piece/ group of pieces, not ordered by number, similarly formatted). Each chapter contains:

- TAGS

- texts with footnotes and bibliographic references

- sticky sidebar with relevant images (scores), recordings (audio and/ or video) so reader can easily access material regardless of where she is in the text

Bibliography

Copyright/ Acknowledgements

About/ Contact

We implemented much of this plan as I intended. But the best solutions were not always obvious from the outset. Throughout our collaboration, seemingly superficial details would provoke deeper questions about user

---

13   By 'writing' I also include the preparation and production of relevant nonverbal media, not only verbal text.

experience. These would then require reflection on my core research, much as my discovery of sticky sidebars had before Reed appeared.

The media gallery layout in the Paths was one significant case. Reed initially proposed a framework based on hideable panels placed to the side of a centered verbal text. One panel contained tags, another contained the media. Readers could open either panel by clicking on its corresponding vertical tab. Tags and media were accessible at any location in the verbal text, as I had requested. We also avoided clutter among multiple categories of information packed into a single screen.

Meanwhile, I was also working on Path 1, 'Seeing the Full Sounding' (described above), and Path 2, 'A Treatise Remix Handbook', which centers on an original radio piece about Cornelius Cardew's graphic score *Treatise* (1967). In both cases, I was in the process of determining how written verbal text would integrate with the audio and video pieces, which contained spoken explanatory verbal texts of their own.

The panel layout was a clean and beautiful option, but observing how the panels separated verbal text and multimedia into discrete zones gave me a slight discomfort. I realized that a certain degree of clutter between words and time-based media was actually desirable. It would help maximize connections between the medial presence of the verbal text and the explanatory potential of the audio and video. This would rhetorically reinforce my exploration of the dialectic of poetry and program in experimental notation as a whole. Thus, we adopted a layout in which tags and media would be permanently visible alongside the main column of verbal text.

Moreover, this reflection shaped the content of Paths 1 and 2. Instead of merely unpacking the nonverbal media, the verbal texts also extend and question them. In Path 1, for example, I discuss sociologist Richard Sennett's concept of 'expressive instructions,' which he elaborates in a discussion of three different recipes for a famous French chicken dish.[14] Although I do not discuss Sennett anywhere in the film, this part of the verbal text links the documentary's specialized subject and filmmaking techniques to practices beyond the film's immediate purview. In Path 2, I cite Cardew's thoughts about *Treatise* both in the verbal text as right-justified, italicized quotes and in the radio piece as spoken text. Each

---

14   Richard Sennett, *The Craftsman* (New Haven: Yale University Press, 2008), pp. 179–93.

context offers the reader a different, sometimes contradictory, framework in which to consider Cardew's theories of notation and improvisation, which changed considerably throughout his career.

Finally, Reed refined my vision for user experience and ultimately opened TP it to a wider public. The website as a whole functions as a sort of meta-score for improvisers. As there is no linear argument from Path to Path, readers can 'choose their own adventure' through the dissertation. I had planned for tags, or Topics, to provide the principal link between Paths. Reed included additional 'tactile' and user-friendly elements for the same purpose. The homepage and main navigation menu, for example, list all the chapters in random order with rollover GIF animations and visual icons, respectively. These entice the reader to interact with the structure more intuitively than my original plan. She also added a search function, with which readers can cut across Paths by effectively inventing and implementing their own index.

Lesson #2: Work with and Learn from Designers

This lesson would seem like an inevitable conclusion to the previous section. However, collaboration is not, as a rule, the center of doctoral training. The task of the PhD student in any discipline, as my supervisors and others such as artist and design scholar Ken Friedman[15] have framed it, is to show that one is capable of independent research by making an original contribution to one's field. Universities award PhDs to *individuals* who accomplish this task. In AR, the medium of presentation may form an inherent part of that original contribution, as it does in TP. So it follows that students in these cases should also perform the design work, just as they write what might more conventionally be called 'content'. Or does it?

RC founder and AR theorist Michael Schwab indirectly shares this line of thought when he cites *RC*'s often 'underdesigned' appearance, a consequence of artists meeting the sometimes unwieldy challenge of integrating graphic and web design as crucial elements of their research:

---

15 Ken Friedman, 'Now that We're Different, What's Still the Same?', in *Doctoral Writing in the Creative and Performing Arts*, ed. by Louise Ravelli, Bruce Paltridge and Sue Starfield (Faringdon: Libri Publishing, 2014), pp. 237–62 (p. 249).

[O]ne can claim that the *RC* allows the calibration of an exposition, where this calibration forms an essential part of the research's experience and meaning. One might also want to add that a sense of integrity may be given space at the experiential core of a researcher's practice. Conversely, one may question the corporate sites of research—including those of academia—for interfering with the meaning of research through the control of the presentation.[16]

Schwab rightly prioritizes the intellectual and experiential role of design in AR. He values *RC* as a way to give AR scholars individual control over their presentation formats, even at the expense of conventional aesthetic standards. But he paints an unnecessarily binary picture of *RC* and the alternatives: do it yourself, or leave it to professional managers or corporate journals and research portals. This view does not take into account how researchers and designers can work *together* to *enhance* AR projects.

My experience with TP shows a fruitful middle way between the poles of total DIY and outsourcing. Collaborating with Reed involved no abdication of intellectual or aesthetic responsibility, nor any semblance of 'interference'. On the contrary, her contribution maintained high professional design standards *and* provided me valuable conceptual feedback, without which the project might not have reached its full potential. Reed's role was more akin to a lab partner than an employee.

Our collaboration is just one example of how designers and PhD students—not only in AR—can collaborate. Designers can advise on visual representation for the web in any discipline and at every scale, from site architecture, layout and typography to graphs and tables. Even when funds are not available for integral web design, students can learn from designers through online tutorials, workshops and other forums. Looking toward the future, one hopes that as native digital publishing becomes increasingly normalized, funding for designers and design education might become a more routine part of doctoral training in general.

---

16   Michael Schwab, 'Expositions in the Research Catelogue', in *The Exposition of Artistic Research Publishing Art in Academia*, ed. by Michael Schwab and Henk Borgdorff (Leiden: Leiden University Press, 2014), pp. 92–104 (p. 100).

Lesson #3: AR and Natively Digital Dissertations are Good for Each Other

In this chapter, I have traced a few important elements and moments in writing TP as a native, media-rich website. Three interrelated concepts— readership, representation and reflection—anchor my narrative, which may be summarized as follows. Firstly, my research topic, notation for improvisers, is of interest to many nonacademics. I adopted the website format principally to reach them. Since my research objects were poorly represented in mainstream publishing and scholarly discourse, it was important to provide readers high-quality documentation of work by other artists, and of my own artistic interventions at the heart of the research process. Secondly, the main task of the web design was to represent artistic documentation appropriately together with lengthy verbal texts. This required making media and texts *simultaneously* accessible, so readers could compare them and perceive the kinds of tension between notation and improvisation I sought to articulate. A layout built around a sticky media gallery made it possible to represent audio, video, scores, and verbal text as dynamic partners in my arguments. Thirdly, personal experimentation with web tools and collaboration with a professional designer not only led to a satisfactory execution of the design strategy. They also clarified and enriched the research process itself by requiring deeper reflection on connections between form and content, verbal text and multimedia, and theory and practice.

I hope that this personal experience outlines some common issues in writing a native digital dissertation and thus helps other PhD students to determine whether, and how, to write their own. Zooming out, I hope these field notes show how AR can contribute to discussions around digital publishing in academia more generally. The case of TP exemplifies Schwab's claim that

> artistic research might offer a point of reference for any form of contemporary research, because an understanding of the impact of the presentation format not only enhances the communicative powers of a research project, but also shapes the research process and is reflected in its findings.[17]

---

17 Michael Schwab, 'The Research Catalogue: A Model for Dissertations and Theses', in *The SAGE Handbook of Digital Dissertations and Theses*, ed. by Richard Andrews et al. (London: SAGE, 2012), pp. 339–54 (p. 339).

Zooming out further still, I hope TP suggests how the marriage of AR and digital publishing might fuel new forms of public intellectualism beyond the orbits of academia and the art world. TP was not only a dissertation to be defended and shelved; it is a living meta-work which continues to engage a wide variety of thinkers, practitioners and even popular media.[18] Though TP's scope is of course limited, it, nonetheless, publicly entangles creativity and critical thought in a medium where they are in shorter and shorter supply. Tomorrow's PhD students would do themselves and the world a favor by redressing that scarcity.

Bibliography

Borgdorff, Henk, 'The Production of Knowledge in Artistic Research', in *The Routledge Companion to Research in the Arts*, ed. by Michael Biggs and Hendrik Karlsson (New York: Routledge, 2011), pp. 44–63.

Cobussen, Marcel, 'Deconstruction in Music' (PhD dissertation, Erasmus University, 2002), http://www.deconstruction-in-music.com/navbar/index.html

Cowley, Julian, 'Choose Your Own Route' (April 2017), *The Wire*, p. 16.

Friedman, Ken, 'Now that We're Different, What's Still the Same?', in *Doctoral Writing in the Creative and Performing Arts*, ed. by Louise Ravelli, Bruce Paltridge and Sue Starfield (Faringdon: Libri Publishing, 2014), pp. 237–62.

Munford, Monty, 'How WordPress Ate the Internet in 2016... And the World in 2017' (December 22, 2016), *Forbes*, https://www.forbes.com/sites/montymunford/2016/12/22/how-wordpress-ate-the-internet-in-2016-and-the-world-in-2017/

Research Catalogue: An International Database for Artistic Research, https://www.researchcatalogue.net/

Schwab, Michael, 'Expositions in the Research Catelogue', in *The Exposition of Artistic Research Publishing Art in Academia*, ed. by Michael Schwab and Henk Borgdorff (Leiden: Leiden University Press, 2014), pp. 92–104.

Schwab, Michael, 'The Research Catalogue: A Model for Dissertations and Theses', in *The SAGE Handbook of Digital Dissertations and Theses*, ed. by Richard Andrews et al. (London: SAGE, 2012), pp. 339–54.

Sennett, Richard, *The Craftsman* (New Haven: Yale University Press, 2008).

Williams, Christopher, 'Tactile Paths: On and through Notation for Improvisers' (PhD dissertation, Leiden University, 2016), http://www.tactilepaths.net/

---

18   Julian Cowley, 'Choose Your Own Route' (April 2017), *The Wire*, p. 16.

List of illustrations

Chapter 2

Chapter 3

Chapter 9

Chapter 12

Chapter 13

Index

About the Team

Alessandra Tosi was the managing editor for this book.

Adèle Kreager performed the copy-editing, proofreading and indexing.

Anna Gatti designed the cover. The cover was produced in InDesign using the Fontin font.

Luca Baffa typeset the book in InDesign and produced the paperback and hardback editions. The text font is Tex Gyre Pagella; the heading font is Californian FB. Luca produced the EPUB, MOBI, PDF, HTML, and XML editions — the conversion is performed with open source software freely available on our GitHub page (https://github.com/OpenBookPublishers).

This book need not end here...

Share

All our books — including the one you have just read — are free to access online so that students, researchers and members of the public who can't afford a printed edition will have access to the same ideas. This title will be accessed online by hundreds of readers each month across the globe: why not share the link so that someone you know is one of them?

This book and additional content is available at:

https://doi.org/10.11647/OBP.0239

Customise

Personalise your copy of this book or design new books using OBP and third-party material. Take chapters or whole books from our published list and make a special edition, a new anthology or an illuminating coursepack. Each customised edition will be produced as a paperback and a downloadable PDF.

Find out more at:

https://www.openbookpublishers.com/section/59/1

You may also be interested in:

Digital Technology and the Practices of Humanities Research

Jennifer Edmond (ed.)

https://doi.org/10.11647/OBP.0192

Engaging Researchers with Data Management

The Cookbook

Connie Clare, Maria Cruz, Elli Papadopoulou, James Savage, Marta Teperek, Yan Wang, Iza Witkowska, and Joanne Yeomans

https://doi.org/10.11647/OBP.0185

Digital Scholarly Editing

Theories and Practices

Matthew James Driscoll and Elena Pierazzo (eds)

https://doi.org/10.11647/OBP.0095

CPSIA information can be obtained
at www.ICGtesting.com
Printed in the USA
LVHW070619150721
692728LV00003B/16